Chasing The Waste Out Of Sales

Conway Management Company

15 Trafalgar Square, Nashua, NH 03063-1989 • Tel: 603-889-1130 • Fax: 603-889-0033
Website: www.conwaymgmt.com

Special thanks to all of the people at Conway Management who helped in the writing, design and production of this book. Their help has been invaluable. We also appreciate the input that clients have shared with us. This book is for you and all of the other individuals and teams out there striving to help their organizations sell more and grow.

Written by: Mary Ellen McNamara
Senior Project Director: Jody Ayers
Cover Design: wedüdesign
Director of Design & Production: Margot von Rosenvinge Horn

Conway Management Company, Inc
15 Trafalgar Square
Nashua, New Hampshire, 03063 USA
603-889-1130 • 800-359-0099
www.conwaymgmt.com

10 9 8 7 6 5 4 3 2

Foreword
by
William E. Conway

This book is about selling more and growing a business profitably. It takes a lot more than just good selling and marketing. It takes improving and optimizing sales, marketing and customer service as well as improvement in new product development, costs, productivity, asset management, R&D, engineering, human relations and more. By identifying, quantifying, prioritizing and eliminating the WASTE of material, capital, time and lost gross margin $ from lost sales, any company can achieve great things. And selling more and growing the business profitably leads to a strong profit and loss statement, balance sheet and cash flow with an excellent return on investment, assets, and sales...in other words, ***a better business***.

This book shows how to transform your business into a high growth, high quality, high profit company. The keys to success are the elements and principles of *The Right Way To Manage©*, the Conway system of *continuous process improvement*. Broad-based improvement requires hard work, working on the right things and the continuous study, change and improvement of the work and work processes throughout the business, including sales and marketing.

A business is a complete system. When key parts of that system are running right, things usually go well. When one hears that a particular company has a great marketing and sales business, you also often find that that company has the best quality products and services. The best companies also have low costs and expenses, great new products, new technology, strong R&D, and excellent asset control.

For example, look at General Electric, Southwest Airlines and IBM, all of which look to be super companies for many years in the future. All three companies appear well-balanced with solid management and excellent products and services. Customers want and need to buy from them at prices they are willing to accept. These companies have relatively low costs and expenses. Southwest already had them and the other two learned that low costs and expenses were what they needed. They have been in excellent growth markets; they learned how to develop the new products and services for growth. They provide excellent products and services; these companies are doing well in difficult market and economic conditions.

This book is a guide to those who want to develop a truly world class operation.

Included in this book are some actions which only general management can take; it requires leadership and commitment. Here are some examples:

- Identify, quantify and prioritize the opportunities for improvement in your company, division or department.
- Decide which business units can benefit the most from a broad transformation to sell more and grow profitably.
- Move into action to run your business more effectively and profitably.
- Make sure that your Sales, Marketing and Customer Service people utilize The Right Way To Manage© in local areas as well as across the board for measured improvements.

At the same time, there are actions which require sales and marketing leadership.

- Learn more about your work to learn how to improve it. Then act as a leader to change what you do as an individual and as a manager.
- Increase the value added work. Work on the right things!
- Collect data on why your major customers and potential customers buy and do not buy from your company. Analyze and initiate action so you sell more and grow profitably.
- Eliminate the troubles, errors and complexities in the work.

The potential for improvement in most companies is far greater than most people can imagine, but it requires bold action and leadership. It requires getting off the sidelines and into the game!

What will the book do?

This book shows you how to apply the proven and highly successful Right Way To Manage© philosophy to the sales process. It shows you how to use simple charting tools to find and eliminate waste in the sales process: waste of time, waste of material, waste of capital, and waste resulting from missed opportunities. It shows you how to provide your organization with customer feedback that can be used to improve your ability to meet customer requirements and consequently grow your sales. And it helps you understand how your organization can remove the waste that results in lost sales or lost gross margin. **It provides you with knowledge, tools and methods so that you can sell more!**

What's in it for you?

Think about it for a minute. How much time do you spend talking to uninterested or unqualified leads? How often do you have to discount prices because of delivery delays or quality issues? How much of your energy is expended expediting orders and handling errors, troubles, problems and complexities? How much money is spent on marketing pieces that end up in the trash or don't generate the expected results? These are all forms of waste—waste that typically is chronic and accepted as "just the way things are".

Now imagine a sales process that has only value added activities. This process would produce sales and customer/consumer information that continuously improves your organization's ability to provide customers with what they want, when they want it, at a price they are willing to pay. Your organization's ability to move with the market in the future would be a predictable outcome of your sales process. You would be working effectively with the rest of your organization to identify and meet customer needs and requirements. You would be calling on more qualified prospects and converting more leads into sales. In short you would be more successful.

What's in it for your company?

> *Most companies working in continuous process improvement start with manufacturing because they believe it is easier. It's an area where numbers are usually readily available, and it seems like an obvious target for quick results. The real litmus test for companies, however, is sales gross margin dollars. When companies tell me they are already strong in continuous improvement, I quickly try to find out whether the sales department is deeply and truly involved. If not, I know they failed the test.*
>
> *—William E. Conway*

The measure of business improvement is really quite simple. You want gross margin to improve. Dollars from sales should increase, while direct costs and expenses decrease. And you want a sustainable system that will ensure you continue to get solid business results. Simple to understand, not so simple to execute.

Continuous improvement enables you to boost gross margin dollars, while simultaneously enhancing the quality of your products and services. It is all about running a better business. Continuous improvement accomplishes this by eliminating the waste that diminishes quality, results in inconsistency, and reduces productivity. Eliminating waste leads to improved quality, consistent results, and enhanced productivity—and enables your organization to meet or exceed customer needs.

Companies often have improvement efforts in manufacturing, administration, finance, customer service, and other departments—but typically fail to include sales and marketing in the improvement process. The assumption is that sales and marketing are "different" from other departments and that the continuous process improvement approach does not apply.

In reality, the largest waste in most commercial and industrial organizations is lost gross margin that results from sales not made, sub-optimal pricing, and excessive costs and unnecessary expenses in and affecting the sales and marketing process.

What is this book about?

The information in this book is based on a time-tested system of management developed by William E. (Bill) Conway while he was CEO of a Fortune 500 company and after working with noted statistician and management expert Dr. W. Edwards Deming. Bill invited Dr. Deming to help him bring the principles of statistical control, collaborative customer/supplier relationships, and quality management to his company.

Bill saw the value of incorporating Dr. Deming's message with his current improvement efforts. He understood that continuous improvement of all work and work processes is the most effective route to providing quality products and services that please external customers. He also understood that this method applied to all processes whether you are making a widget, a memory chip, designing a new product, planning a product launch or making a sale.

Bill called the new system The Right Way To Manage©. It is a system for the continual and relentless study, change, and improvement of all work and work processes.

How to use this handbook

The information in this handbook is designed to help you apply continuous improvement concepts to the work of sales, enabling you to boost sales and increase gross margin dollars.

The handbook is divided into three sections. The first covers the basic concepts and principles you should be familiar with before getting started. The second outlines how to search out and eliminate waste. The third describes data-gathering tools and charting techniques to be used as you implement continuous process improvement.

Although your entire department and organization including top management should be involved in the drive to eliminate waste, this book will help you learn what you can do as an individual. Everyone should be part of the continuous improvement process, with management leading the charge, but you don't need to wait until everyone's on board to begin doing your part.

After completing this book, you'll know how to recognize waste, uncover it, quantify its cost to your organization, and eliminate it for long-lasting improvements.

Section 1

Concepts & Principles

Work & Waste

Key points in this chapter:

- The definition of waste
- All waste comes from the work
- How big the waste is

In the Right Way To Manage© system, we define waste as **the difference between the way things are now and the way things could or should be if everything were right.** We also believe that all of the waste comes from the work, either not working on the right things or not doing things the best possible way, all of which contributes to reduced sales and grow margin dollars. How much more could you sell if you and others never had to spend time handling troubles, problems, errors and complexities?

Our experience shows that the waste in any organization is 20% - 50% of gross revenues! If you are like most people, you probably find it difficult to believe that waste represents such a large percentage of an organization's gross revenues. Most people think it is only around 5%. That is because most of the waste is chronic and systemic. It is built into the way departments and organizations are run, the way products and services are sold, the way people do their work. It has become and accepted way of doing business and people no longer see it as waste.

When you truly believe that waste constitutes a high percentage of gross revenues, you become dissatisfied with the way things are, recognize that there are enormous opportunities for improvement and work vigorously to change the status quo. Realizing the magnitude of the waste is a key to motivating change.

So how do you go after all of that waste?

The following Model provides you with a helpful framework.

The Model

Key points in this chapter:

- Your job in continuous improvement
- An explanation of the Model for Continuous Process Improvement
- Waste from lost sales, lost profits and high costs and expenses equals at least 20% - 50%
- The concept of "process"
- The power of quantifying waste

Before we continue, think a minute and ask yourself: *What is your job? What are the key activities you do to accomplish your job? Is it "close the deal"? Make money? Increase sales? Find more customers? Make it easier for people to sell?* Whatever it is, write it down.

Now let's talk about what your job is when you are working in a system of continuous improvement. You really have two jobs: 1) do the job to the best of your ability, and 2) continually study, change and improve your work and work processes. The Model for Continuous Improvement shows you the basic elements of the Right Way To Manage© and how they fit together. The area outlined in red is your "key" to success.

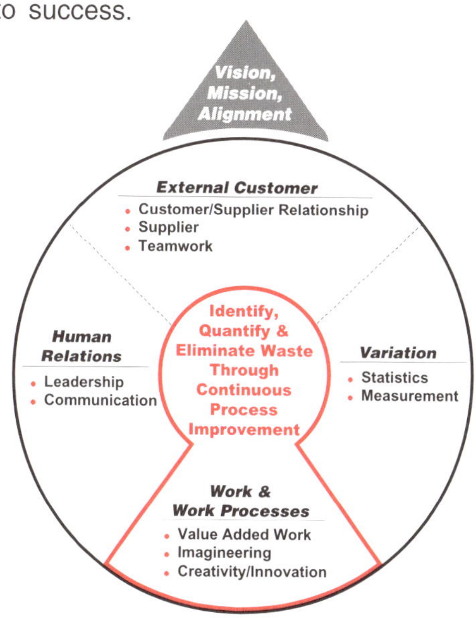

Notice that the Model is organized around a center circle labeled "Identify, quantify and eliminate waste through continuous process improvement". We refer to this as "the core activity" and it should be a key part of your everyday activities. What this means is you should constantly be searching for waste, determining how big it is, selecting the right things to work on and then getting rid of the waste by continuously improving the work processes. Done properly, the core activity helps people to find and work on the right things to grow and improve the business.

Surrounding the core activity are four supporting elements that contain concepts and tools an organization can use to eliminate waste and improve the quality of all work, including marketing and sales. The 4 supporting elements are the external customer, variation, work/work process, and human relations. Vision, mission and alignment provide direction and focus. This book describes each of these in detail.

The Power Of Quantification

Quantification is a powerful tool that enables you to see just how huge the waste is and gives you the drive necessary to eliminate it. Quantification means putting a dollar value on how much the waste is costing your organization on an annual basis. Numbers and dollars are the language of business and management. Quantification allows everyone to speak the same language and have a common frame of reference when comparing opportunities. It also ensures that everyone stays focused on running a better business. Once you have identified the waste—for example, lost business due to unacceptable delivery times—you can determine how often that happens, how much gross margin opportunity is lost, and determine how much profit is lost annually as a result.

Once the waste associated with sales and marketing has been quantified, you can decide what to work on and select a process to improve. This is an extremely important point: *quantify the waste before selecting improvement projects.* This keeps you focused on the potential business impact of your efforts. In many organizations, people pick a project and then look for the waste. This often results in working on the wrong thing, which can be the biggest waste of all.

The Value Of Quantifying

Waste occurs even in well-run operations. Waste is so embedded in the work and work processes, most people fail to recognize it. Only by questioning everything about the work, what is done and how it is done, do people really begin to see all the opportunities for improvement.

During one Conway Management workshop presented to a supermarket chain, the owner-president was extremely hostile to the idea that his company had 35% waste. The company was considered one of the best-run in the country, and he found the use of the word "waste" insulting. During the workshop, employees were understandably reluctant to respond when asked how much time the average person wasted per day in the organization.

One of the workshop leaders started asking a few pointed questions. "Do you folks have to wait for anything around here?" "We have to wait for deliveries," a brave soul answered. In 10 minutes, they compiled a list of things they waited for.

"OK, maybe now you can answer my question. How much time do you think the average person wastes per hour in your organization?" Finally someone responded, "Twenty minutes." The workshop leader asked if everyone agreed and they nodded. "To be conservative, let's cut that in half. We'll say ten minutes an hour." He then calculated how much that would total over the course of year, based on the average hourly rate for a typical eight-hour day. The answer: as much as $5 million.

Then the Senior Associate moved on to another topic: how many potential customers could or should shop at this supermarket chain but did not. After a few minutes of silence, he asked, "What are some reasons customers don't select your store?" The group quickly listed ten reasons and agreed that if they eliminated those reasons, they could probably gain an additional 10 customers an hour. Calculating the average annual revenue per customer and cutting that number in half as well yielded another $3 million.

By this time the president was smiling. He stood up and said: "All right, all right, you can say it. Now show us how to get rid of it."

In order for him to understand the concept of waste, he needed help. That help was quantification.

The Concept Of Process

A basic tenet of The Right Way To Manage© is that all work is part of a process. A process is a series of steps or activities required to take inputs of various types, add value to them, and turn them into outputs. You may not be accustomed to thinking about your work in this way. The tendency is to view each task as a discrete event. Typical "to-do" lists reflect this: Call Account A; Check on order for Customer X; Fill out call report; Submit expense report; etc. Each one of these tasks is part of a larger process. For example, "call customer A" may be part of the lead conversion process. Checking on an order for Customer X may be part of the order fulfillment process.

Recognizing that what you do is part of a process enables you to start looking at work in an entirely different light. As you start examining processes you begin to understand how interdependent you are with others to get the work done. You can also quickly see all the places where things can go wrong, why things take so long, and you begin to question, why do we do things this way?

Common wisdom has been that if people would just do their job or tasks correctly, everything would always turn out well. In reality, individuals can be performing their jobs precisely as specified, but if the process has built-in problems, or people don't understand customer requirements or how what they do affects others in the process, errors and waste will naturally occur. In fact, more than 90% of the time, problems are caused by waste in a process or system—and not because individuals are not trying to do the right thing.

For example, as an order flows through your organization, it requires action by other departments or systems. As you well know, this process is prone to opportunities for errors at every turn. Incorrect or incomplete information can create problems resulting in waste and potential customer dissatisfaction. It often becomes your job to untangle the mess. When you check into the situation, you often find that people were doing the best they could within a flawed system.

Sales departments are usually more closely connected to the external customer than the rest of the organization, which makes them a vital source of information about customer's needs, wants, requirements and levels of satisfaction. However, to translate customer needs and requirements into products and orders, sales must work together with other parts of the organization through systems and processes. It is often at these hand-offs or interfaces between people and steps in a process that confusion and problems occur leading to waste of all kinds.

It is precisely for these reasons that process improvement, not just task improvement, is key to success.

Processes In Sales

The sales and marketing department, is part of a system that is comprised of many processes. Some of the processes found in sales departments are:

- Generating leads
- Converting leads
- Setting up a sales call
- Making a sales call
- Preparing new promotional materials
- Conducting a demonstration
- Completing a proposal
- Replying to requests for information
- Getting the order
- Placing an order
- Growing accounts
- Telemarketing
- Training sales people
- Measuring and improving customer satisfaction

What are some additional processes in your department?

1.

2.

3.

4.

5.

What works well about these processes?

What goes wrong with the processes?

This is just the start of some of the data you will want to collect to begin to study your processes.

Making The Connections

To avoid creating process conflicts, sales people need to know as much as possible about how their organization delivers on the products and services sold.

> A Conway Senior Associate recalls working in sales for a hotel in Kansas. February weekends were a particularly difficult time to attract people to the hotel. She was thrilled when she booked The Cattlemen's Association, prize-winning cattle and all, for a Friday through Sunday afternoon in February, and then topped that off with a 350-person Bar Mitzvah for Sunday night. Needless to say, the banquet staff was not pleased to learn they had one hour to turn the ballroom from a cattle show to a first-class Bar Mitzvah! She was "just doing her job". But as Dr. Deming said, "Just doing your job is not enough. It never has been."

This example illustrates why people must understand all of the steps in the process to understand the impact of their piece. These connections are the interfaces and interstices of a process. Interfaces are the meeting points between people; interstices, the meeting points between things. It is at these connections that major sources of waste occur such as missed hand-offs, delays, miscommunication, transit, reconciliation, and rework of all kinds. Because work is a series of interrelated steps that do not exist alone in departments or silos, true improvements cannot be made unless processes are improved.

A Way To Look At Sales Processes

Listed below are block flow charts for some of the sales processes previously identified. Draw a block flow chart for all the steps you can think of in these processes in your organization. Add as much detail as you can.

Generating Leads

Converting Leads

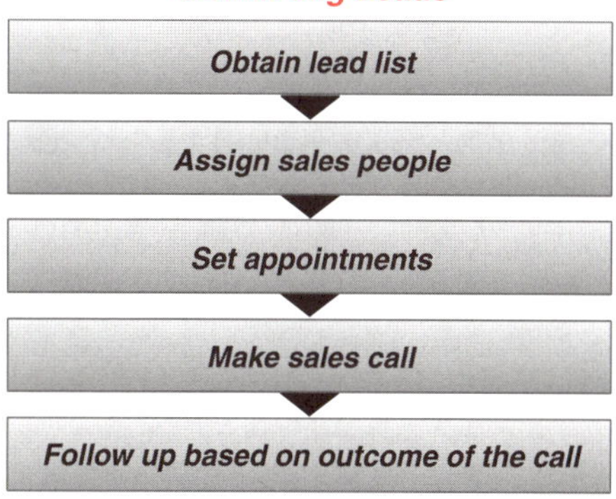

Making A Sales Call

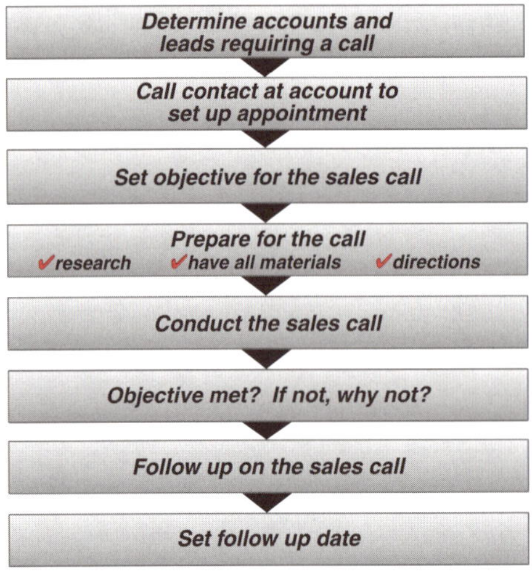

For the flow chart you drew, identify what can and does go wrong at each step and at the hand-offs between steps, particularly steps that involve different people or different departments.

For each thing that could go wrong, try to list the consequences (the extra work steps) that occur and how much additional time and cost that adds. If you multiply this by the number of occurrences per year, the dollar figure is often staggering! Not to mention the potential customer dissatisfaction that results.

In addition, identify what must happen for each step to go "right" every time.

It is easy to see that you could focus on the most important opportunities for improvement by collecting data and prioritizing things that go wrong. By studying the process and the root causes of the problems, you can reduce or eliminate them. By finding a way to incorporate what must happen for the step to go "right", you can prevent problems from occurring. Obviously this requires a team effort by everyone involved in the process.

The Vision/Mission & Alignment

Key points in this chapter:

- Why vision, mission and alignment are so important to the sales department
- How to determine the vital few objectives for your department and organization
- How to use the vital few objectives to align the organization

Most organizations today use mission and vision statements as a means to articulate what they are in business to do and what they aspire to become. The value of these statements lies in their power to provide meaningful purpose for everyone's work and as guidance to make decisions every day about what is important to work on. Just as geese flying in formation operate with a few clear rules to guide their collective behavior, your organization's vision, mission and values serve as powerful guides for organizational effectiveness.

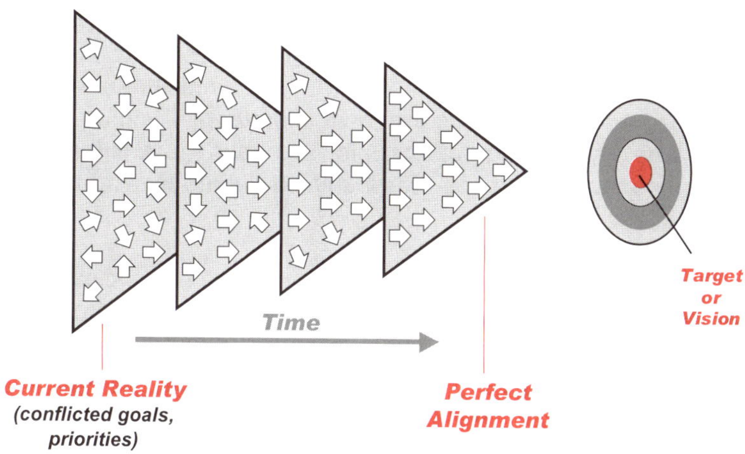

To get an idea of vision, mission and alignment in action, think of a time when you and a group of people set out to accomplish something that was really challenging and meaningful. It could have been anything from building a tree house to building a business. You all shared the same vision of the outcome; you agreed on how each of you would contribute and how you would work together; you clarified and resolved issues quickly, and you worked passionately to get the job done. Imagine what your organization or department would be like if everyone had the same picture of where you were going, what the most important things were to work on both individually and collectively and you had great communication and feedback systems that helped to see your progress and adjust course as necessary. You all knew there was something meaningful in it for you. You would not have to tell everyone what to do every day; people would be guided by the vision and the mission. How do organizations go about creating that kind of focused and flawless execution?

There are varying definitions for mission and vision. The following captures their essence in a few words:

Vision: a powerful, inspiring mental image of a desirable achievable future state.

Mission: a clearly defined purpose or core reason for being.

An interesting exercise at your next sales meeting would be to ask everyone to write down what they believe the purpose of the department is, what the key goals are and the essential actions that must be taken to accomplish them. Then ask people to share what they have written. Does everyone have the same picture of where you are headed and what actions are most important? If you are typical, there will be a wide range of answers. The good news is that these can become the starting point for a spirited debate leading to a clear plan.

The first challenge is to make sure that everyone has a clear understanding of what you are in business to do and what noble purpose that holds. In other words, what is it about the work that you do that makes it worthwhile to get up every day and do your best? This mission or meaningful purpose attracts and keeps great performers. For example, Lexus' mission is "the relentless pursuit of perfection." No matter what your job is at Lexus, you know you are part of trying to perfect every product and service. As a salesperson, you are not just selling a car, you are selling "perfection," both in the product and in the sales and ownership experience.

The next challenge is to ensure that everyone shares a common picture of where you are going or what you are trying to become. When Lexus first entered the market, their vision was to "beat Benz".

For years, Mercedes Benz had been the leader in the luxury automobile market, so beating them was a clear and measurable picture of Lexus' desired future state. Knowing the vision allows everyone to determine what actions to take to get there. Similarly, a vision of what you want your sales department to look like or how you want it to operate provides a focus for planning and measuring progress.

Once you know the mission and vision, you are ready to make a plan. You'll need to determine your strategy for achieving your vision. Once you do, you will be ready to identify the vital few objectives essential to achieving your vision. There are always many things that can be done to accomplish your goal, but only a few of which are the real drivers for success. Organizations that zealously focus on three to five key objectives are consistently more effective than those who spread their efforts and resources too broadly. So, the vital few must be:

- chosen carefully,
- communicated to everyone,
- have high priority and visibility, and
- be vigorously tracked for progress.

Used with purposeful discipline, the simple questions outlined below can get you headed in the right direction.

A Look At Your Vision & Mission

- What is your organization's vision?
- What is your organization's mission?
- What are your organization's Vital Few Objectives?
- What is your department's vision? How strongly does it support the organization's vision? With whom must you collaborate to achieve your vision?
- What is your department's mission? How strongly does it support the organization's mission? With whom must you collaborate to achieve your mission?

- What are your department's Vital Few Objectives? How strongly do they support those of the organization? With whom must you collaborate to achieve your vital few objectives?
- How do these influence what you work on every day?
- On a scale of 1(low)-10 (high), how would you rate your organizational alignment?
- On a scale of 1 – 10, how would you rate your departmental alignment?
- What specific waste does a lack of alignment cause you and the organization?
- What processes need to be improved or actions to be taken to enhance alignment within your circle of influence?

Translating The Mission/Vision To Action

How can you take the vision and mission off the wall and translate them into actions and behaviors to achieve them? Begin by asking: What are the vital few objectives or key things at which the organization must excel to attain the vision and mission? List these in the spaces on the matrix diagram below. Now ask: what are the vital few objectives or key things at which sales must excel in order to help the organization achieve its vital few objectives and move toward its vision? List these in the appropriate spaces below.

Now, to check for alignment, ask the question: How strongly does each sales objective support each organizational objective? Place an "S" for strongly, an "M" for moderately, a "W" for weakly and a "O" for not at all in each block. Then look at the pattern. Are all of the organization's objectives being supported by sales? If not, do you need to add some important items? Do some of your objectives in sales not support any of the organization's objectives? If so, should you still be working on them? This tool helps make visible the linkages level by level down through the organization.

Vital Few Objectives For Dept/Area	Vital Few Objectives For Organization			
	1	2	3	4
1				
2				
3				
4				
5				

S = Strong M = Moderate W = Weak O = None

Activity/Task	Vital Few Objectives For Dept/Area				
	1	2	3	4	5
A					
B					
C					
D					
E					
F					

S = Strong M = Moderate W = Weak O = None

To take this thinking to the individual level, list your understanding of your vital few objectives and evaluate how well aligned they are with the vital few objectives for sales. Are you working on things that will enable sales to achieve its goals and objectives? Will all of sales' objectives be accomplished if everyone accomplishes their individual objectives? If the answer to either of these questions is "no", decide what objectives or actions should be added or eliminated. Again, this is a valuable tool for assessing alignment. Do the same for each project of your group.

Another important question to ask is "How much time are you actually spending on activities that support the vital few objectives?" If the answer is "not much", then you need to redirect your efforts.

If you are a manager, use these charts to periodically check for alignment between individual, group and organizational objectives to facilitate effective teamwork and movement toward achieving your vision. You will also need to monitor and measure how much time people are devoting to the vital few.

The Customer/Supplier Relationship: Understanding Your Customers

Key points in this chapter:

- What the customer/supplier relationship is
- The two kinds of customers
- Determining customer requirements

Satisfied and delighted customers are the lifeblood of any organization. Work is fun and rewarding when customers love what you do or provide for them. Providing customers with the highest quality goods or services at the best possible price starts with clearly understanding the customers' needs and requirements and then designing and implementing processes that consistently deliver value.

There are two types of customers: external customers and internal customers. Both are important and have needs that must be met. External customers are the people who pay for your products and services. As Dr. Deming said: "No customers, no orders, no jobs!" Paying attention to the external customers' requirements is essential and focuses the entire organization on doing value added work.

To effectively meet the external customers' needs, you must often work with many internal customers. Understanding and meeting your internal customers' needs and requirements helps the process of producing your product or service to flow smoothly, be problem-free and deliver the highest quality at the lowest total cost.

This customer-supplier relationship extends to your external suppliers as well. From your external customer's point of view, you are responsible for what they buy from you; your suppliers are part of your system. It is increasingly important to build strong supplier partnerships that ensure that you get exactly what you need, in the right quantity, at the right price to be able to meet your customers' needs.

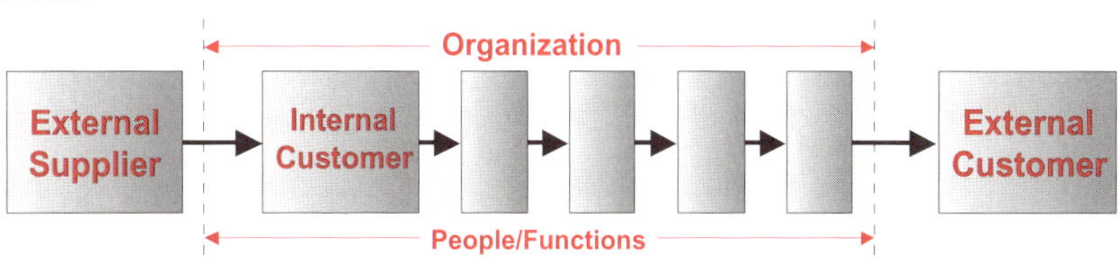

The effectiveness and efficiency of this chain of customer-supplier relationships is key to your organization's success.

The Process-Customer/Supplier Connection

All Work Is Part Of A Process

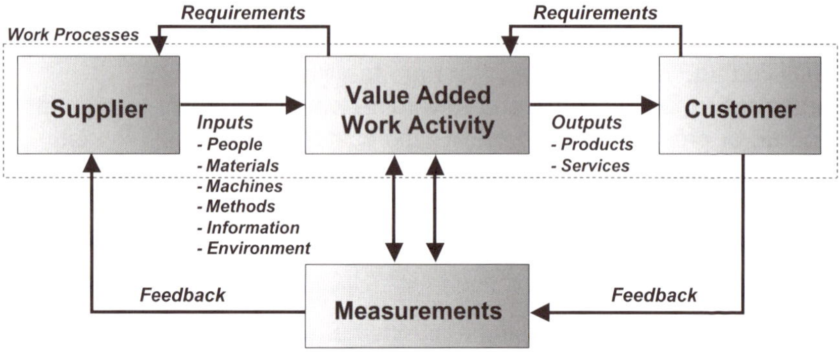

Recognizing that all work is part of a process is central to the concept of customer/supplier relationship.

The Relationship Between Sales & The External Customer

A critical role of sales and marketing departments is to obtain accurate information from the customer and the market—and communicate it effectively to the rest of the organization. Once your organization has this information, processes must be designed and managed so that they are capable of meeting customer requirements. If a customer requires delivery within three days and your system is capable of delivering goods in a week at best, you will be unable to meet a key customer requirement. Your processes must be improved to reduce delivery time.

Determining Customer Requirements

Accurately gauging customer requirements is no easy task. First of all, customers may have trouble clearly articulating their requirements. Secondly, the salesperson has the challenge of interpreting the requirements and translating them into specifications. The specifications are then handed off to others in the organization, with high potential for misunderstandings and misinterpretations in the process. It is easy to see why failure to get customer requirements right causes waste throughout the organization, potentially all the way back to what is purchased from external suppliers.

Ask — Don't Assume!

Making assumptions about what requirements are important to the customer can cause waste. One Conway Management client, a large manufacturing company, went through a difficult, time-consuming process to match paint colors on their deck chairs. The white of the plastic for the seat covers was often a different shade than the white tubing used to make the frame. The company used a special instrument to scrutinize the colors on a finished chair. A high percentage of the chairs were rejected because they did not meet the specifications developed in engineering. When finally asked how important this requirement was, a customer replied: "We don't care about the variation in shades of white at all. We just want the legs to be uniform so the chair does not rock." This was something the supplier didn't even test for! Sales and engineering must work closely with the customer to ensure there is agreement on the requirements and what is most important to customers.

What questions could you ask your customers to gain a better understanding of their needs and requirements? Imagineering can be effective when dealing with a customer who is having a hard time articulating requirements. Imagineering means visualizing what things would be like if everything were right. Ask the customer to imagine that it is the future and he or she is looking back at the sale being discussed. Then ask him or her to explain what would have happened if everything went perfectly with all aspects of the service or product. What would the quality be like? the delivery? the packaging? the invoicing? the supporting documentation? product performance? response time to requests? Notice that customers' requirements extend far beyond simply the "specs" to encompass all aspects of the sales experience.

Another way to determine customer requirements is to know your competition. Customers "vote" with their wallets. Understanding what customers actually buy and why provides insight into features or requirements that customers have not otherwise articulated.

The Relationship Between Sales & Internal Customers

The concept of the customer/supplier relationship encompasses internal customers and suppliers as well. This becomes obvious once you understand the concept of a process. The people doing the work that precedes the work you do, are your suppliers. Those who receive or are affected by your work are customers. So, for everything you do, you are the customer of the previous activity or process and a supplier to the next. Your suppliers and customers may be inside and outside your department, as well as inside and outside your organization.

Who are your internal customers? Be specific as you think about this. Pick a process in which you are involved and try to answer these questions:

- What do you produce?
- Who does your work go to?
- How do your customers judge the quality of your work?
- How well are you meeting their requirements? How do you know?
- What measures do you have to tell you?
- How does the quality of the work you do for your internal customers affect the quality, total cost and delivery to external customers?

Failure to understand or meet the requirements of internal customers can ultimately lead to customer dissatisfaction, lost orders and lost gross margin.

Internal and external customers may have conflicting requirements. For example, each external customer may want specific modifications to your product, such as color or features. If you are to truly satisfy every customer, you will sell them exactly what they want. Internal customers may be focused on improving throughput and reducing costs. Over time, unless you have exceptionally robust processes, you may find that the cost to produce so many variations of your product may drive your costs to an unacceptable level. At this point you will need to make business decisions about the best way to meet customers' needs. Because external customers are the reasons internal customers exist, the focus must be on meeting the needs of external customers while being able to stay in business for the long term. Often, confronting these issues helps to sort out essential customer requirements or "must's" from the "wants" or "nice-to-have's".

Keep in mind that customers can't tell you everything. For example, customers can't tell you that your costs are too high. Low margins will tell you that—unless you're able to pass those costs on to customers. But if you are passing waste in the form of high costs on to your customers, your prices may be too high and this will ultimately cost you sales.

Feedback Mechanisms

Feedback systems and measurement are essential for business success, improvement and organizational learning. A key part of your job is to provide your organization with customer and market feedback that can be used to improve your ability to meet customer requirements. If you had better products and services to market, support, and sell, how much easier would your job be? What would happen to sales? If your deliveries were always on time, with the right quantity and the right merchandise, how would that help you do a better job?

Many organizations lack effective mechanisms for gathering useful customer feedback and translating it into information that can be used to improve the system. Customers provide feedback in a number of ways. They might send the product back or complain. Or they might not say anything, but take their business elsewhere. That is why it is critical to have measures that will tell you how well you are meeting customer requirements as well as indicators that will predict how well you will meet them in the future.

Start by asking yourself these four questions:

- Who are your customers? List them by name, revenue, item, or service purchased.
- What are their requirements? List why they buy. Be specific. Also list what they use the product or service for.

- How well are you meeting their requirements?
- How do you know you're meeting their requirements?

To answer the last two questions, you need an ongoing performance feedback system with measurements. Some of the measures will be results measures, which let you know what happened. These might include number and dollars of lost orders, percent on time delivery, number of customer complaints re: poor quality, wrong pricing, etc. These are useful for understanding what is wrong and what needs improvement but they are lagging indicators. Process measures such as cycle time to process an order or percent of leads converted to proposals are valuable predictors of process performance. If they are moving in the right direction, you can predict that your outcomes will improve. Measures which give early indications of where the market is going or how successful your product or company will be are leading indicators. These can include percent of sales from new products or new product development time. Having a well developed set of measures and a system for using them to assess and improve the business is essential to the success of your organization and especially to sales.

Run charts are an excellent tool for measuring process over time. A dashboard of key run charts gives everyone information at a glance about the performance of the department or business.

Systems or processes are needed that capture customer feedback and translate it into measurements that lead to improvement. Many sales departments have measures and data they collect, but they lack systems for using the information to make improvements or they lack the discipline to use the information on a regular basis.

Feedback systems will vary depending on what you seek to improve. Customer surveys, comment cards, and focus groups offer ways to collect performance feedback directly from customers. In addition, you need internal mechanisms to let you know how you're faring with your customers. These can include:

- Lost business summary
- Customer complaint analysis
- Competitive analysis
- Market share statistics
- On-time delivery records
- Specific attributes that show process capability

Getting Started: Baseline Information

Some of the questions facing you as you embark on a course of continuous improvement are:

- What are the 3-5 "must-have" requirements for your top customers?

  As requirements are identified, it helps to understand their relative importance to the customer. What requirements does the customer consider "musts"? What are "wants"? Keep in mind that customer requirements are constantly changing. Yesterday's bells and whistles may become tomorrow's "musts".

  Know Your Customer

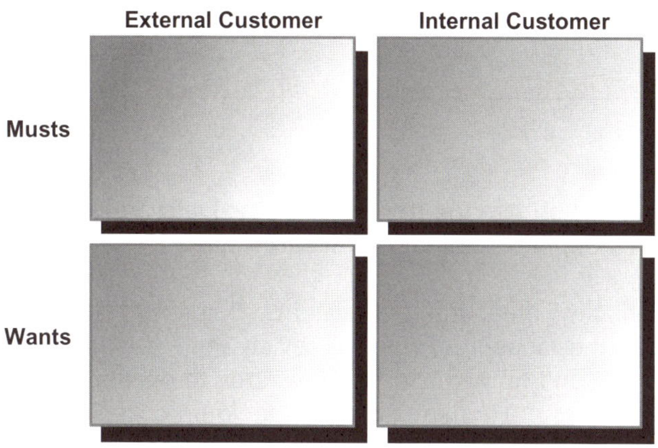

  Use this chart to fill in the "musts" and "wants" of your internal and external customers.

- What measures could you track to determine how well you are meeting the customer's requirements?

  Every process has one or more key measures that tell you how well it is performing and whether you are improving. Because measuring can be expensive and time-consuming, choose measurements carefully.

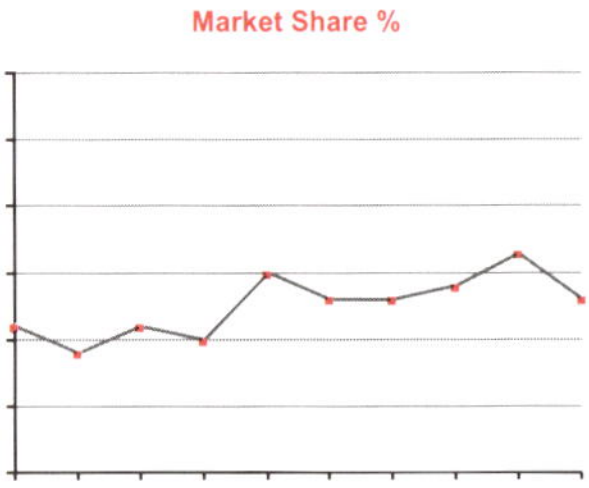

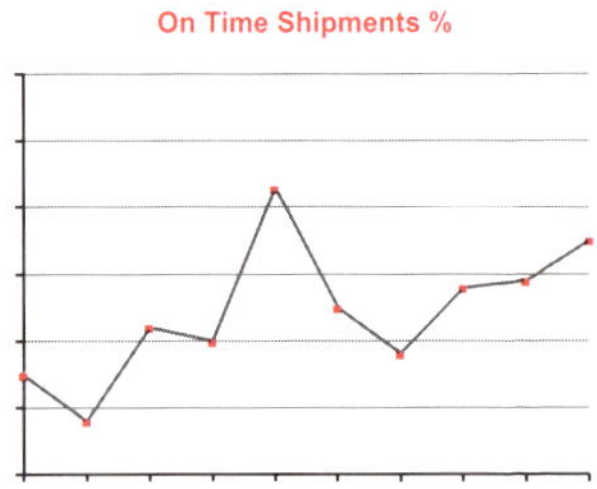

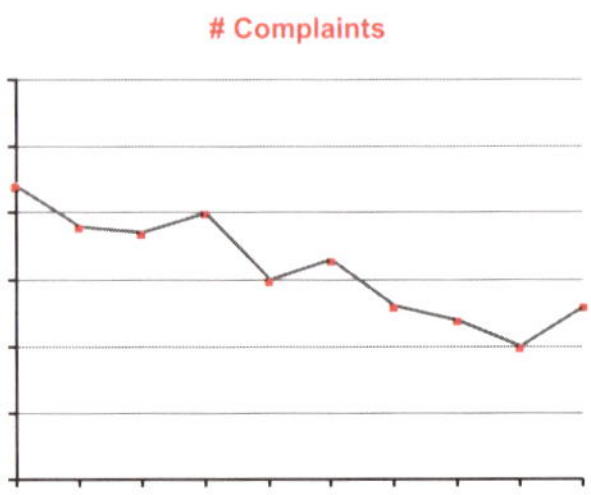

- How can you use the measures to evaluate performance, identify opportunities and eliminate waste?

  An effective measurement system will specify how frequently to review the measures, who will review the measures, who will be involved and how to use the measures to decide on actions for improvement. Remember, data should be collected as a **basis for action.**

Variation

Key points in this chapter:

- Understanding variation's impact on sales
- The high cost of variation
- The relation between customer requirements, process capability and waste
- The two types of variation
- Why mass inspection does not ensure quality

This chapter is one of the most important chapters in this book because the concept of variation underlies everything in improvement. Variation and the use of simple statistics are keys to understanding and improving work systems and processes. Variation is the technical tool used to:

- see the difference between things,
- describe the differences,
- identify waste and
- predict future events.

But you don't need to become a statistician to put these principles to work for you. You only need a basic understanding of variation and how it can help you pinpoint waste and reduce it.

Variation is inherent in everything - people, processes, materials, information, etc. No two things are exactly alike. If you have measuring instruments that are sensitive enough, you can determine the differences. But because no two customers, two deliveries, or no two orders are exactly alike, it is challenging to deliver predictable outcomes for customers and the organization.

Predictability is at the heart of what customers and organizations want and expect. But all you have to do is look back at your process flow charts to see all of the opportunities for variation at every step. Your job is to understand the variation in your products and services as well as the variation in the processes that produced them. The next step is to work with others to manage the variation in the process to improve the results.

The High Cost of Variation

The cost of variation due to input from suppliers is huge. A study at a sneaker manufacturer's plant concluded that over 60% of the work was rework that resulted from variation in the rubber. As an organic substance, rubber has many variables. Because the manufacturer didn't know the range for specific attributes, it accepted all rubber and then had to check the attributes and make production adjustments at every step of the manufacturing process. If instead they knew what the acceptable level of variation in the rubber was, they could use that information to improve the quality of the incoming material, thereby eliminating much of the work and cutting their production time and cost in half.

What waste occurs in your sales process because of variation?

(For example, variation in delivery time, proposal turnaround time, etc.)

A fundamental skill of The Right Way To Manage© is using the information in variation to understand and improve your process.

A run chart or a histogram can be used to show the level at which a process is operating and the amount of variation around that level. You can then use this information to compare performance with customer specifications. Armed with this information, you can then predict with relative certainty what percentage of goods or services can be delivered to the customer as represented in your sales presentation.

Specifications represent “the voice of the customer” — their expectations and requirements. Process capability represents the “voice of the process” — the performance the process actually delivers. When the process always produces results that fit within customer specifications, it is “capable”.

For example, if your organization has a policy that all customer inquiries will receive an appropriate response within 24 hours, that is a specification. If the response time varies between 0 and 24 hours as shown in this chart, your process is capable.

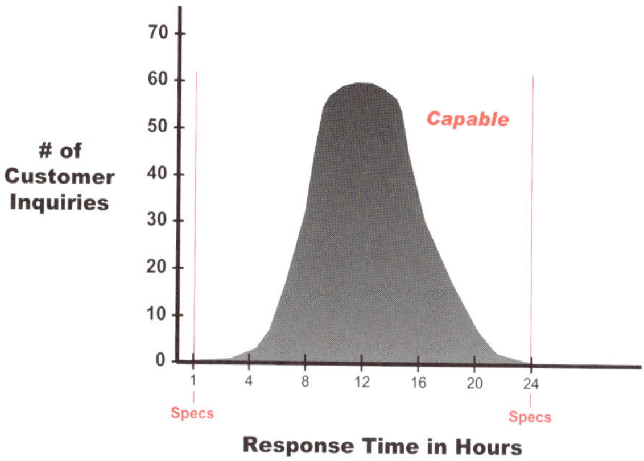

If however, response time varied between 0 and 38 hours as shown in this chart, the process is not capable of meeting organizational requirements.

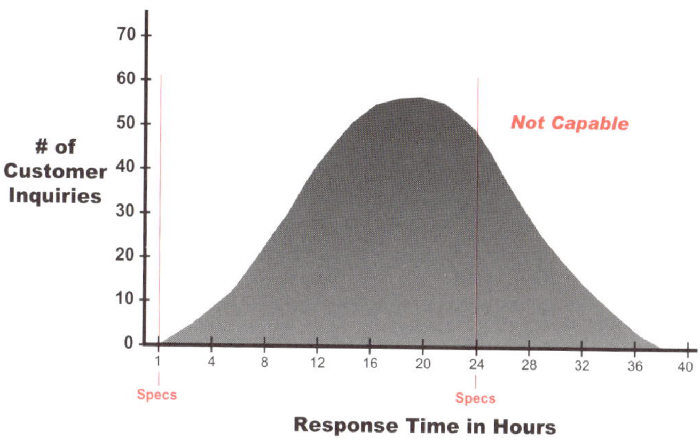

When you make sales promises that exceed the organization's process capability, you often have to compensate with expediting, rework or overtime and ultimately higher costs and lower gross margins. The true cost of your product or service includes these "hidden wastes". Studying, changing, and improving the work will allow you to better meet customer requirements without incurring excess costs.

Improvement Through Knowledge

"The central problem in management, leadership and production is failure to understand the nature and interpretation of variation.

Dr. Lloyd S. Nelson
Former Director of Statistical Methods, Nashua Corporation

A favorite anecdote found in Donald J. Wheeler's book, *Understanding Variation* confirms Dr. Nelson's point..

Many years ago, David Chambers found the following time series on the wall of the office of the president of a shoe company. Here was a simple and powerful presentation of data in context.

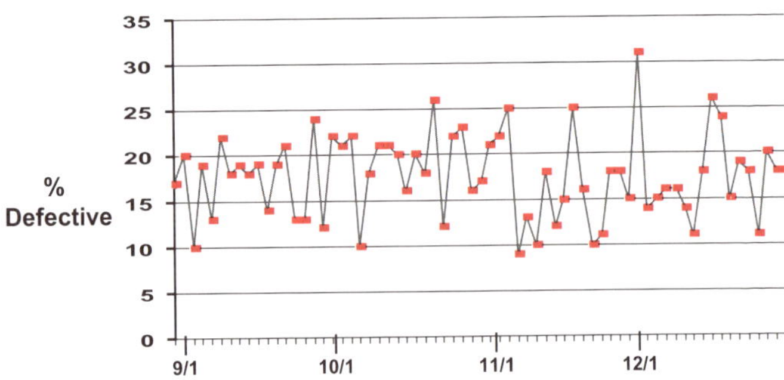

Daily Percentage of Defective Pairs

Intrigued, David asked the president why he had the graph on the wall. The president condescendingly replied that he had the chart on the wall so he could tell how the plant was doing. David immediately responded with: "Tell me how you're doing."

Evidently no one had had the temerity to ask the president this because he paused, looked at the chart on the wall, and said, "Well, some days are better than others!"

The company president did not have a way to analyze and interpret these values and take action for improvement.

The run chart simply showed current reality. The improvement questions are:

- What is the cost of the current level?
- At what level should we be operating?
- If everything were right, how many defective shoes should we be making per day?

If you decide the answer is zero, then the next question is:

- What are the reasons for defective shoes?

The answer invites you to delve into the work to make changes in the materials, equipment, methods and information or other variables. Simply asking people to "try harder" will not do it.

The typical response to variation is to build in contingencies. Building in contingencies (extra lead times, work arounds, expediting, and overspecification for parts) creates waste in the form of excess money, time and capital. A better approach is to use variation to analyze the work so you can take action to reduce the waste.

Sales people may not be able to directly reduce defects, improve delivery time or improve order accuracy. But collecting the facts and data and displaying them in simple charts can provide valuable information to manufacturing, engineering, and product design enabling you to work with them to identify and permanently fix problems.

Random vs. Special Variation

When a customer calls to complain about a late delivery, someone needs to track down the order and find out what happened. Your goal is to resolve the current problem and to take steps to ensure that it doesn't recur. As you know, the variation is likely the result of your system and its variation. However, in order to know what the appropriate action is for improvement, you need to understand what type of variation caused the problem.

There are two types of variation, random and special cause. Random, or common cause variation is due to factors inherent in the way the work is designed. For example, variation in the effectiveness of the sales process may be due to differences in hiring, training methods, experience or product knowledge. If the variation is due to random causes, the cure requires changing the work. In this case, it means changing the way you hire, train, provide product knowledge or methods used to do the work.

Take delivery time, for example. If your delivery time *averages* 7 days and you promise 7 days, some of the time you will be on time and some of the time you won't. So, when the shipment does not arrive in 7 days, it would be a waste of everyone's time to run around and look for reasons why. If you want to ensure 7 days, you will need to study the delivery process to:

- understand all of the steps, how long each step takes and the time between steps
- identify causes of delays and long cycle times
- make changes to the way the work is done so that the average and the maximum are below 7 days

Improving the process requires changes to reduce the average time as well as the range of possible delivery times.

Special or assignable cause variation is due to an identifiable event, something that is truly unusual or different. In this case, it is worthwhile to track down the root cause of a problem and eliminate it. Using the delivery example, if a shipment took 15 days and data shows your process capability is 6-7 days, analysis of the data may indicate that something unusual happened. Investigate. Find out what the special cause was and take appropriate action to prevent its recurrence. Perhaps the delivery was late because of bad weather conditions or a computer failure. This does not require making system changes, such as deciding that all deliveries will now go through a different vendor or different shipping point. It may require implementing procedures for when the special events recur. Changing the system to address special causes is known as tampering and can make things worse. (The tool for distinguishing between special and common cause variation is the control chart, which is explained in more detail in Section 3.)

You may not have statistical process control data about key parameters of your products or services, but if you did, that could improve your ability to sell your product or service to the right customer at the right price. Many companies now require a statistical control chart along with a Certificate of Assurance from a supplier in order to make better buying decisions. If your process is capable with a high level of statistical control, that is a competitive advantage you can sell.

The High Cost Of Inspection

Organizations turn to inspection because they know their processes produce errors or defects and inspection increases their confidence that what they are sending to the customer is "good". When you have someone review or check a shipment, you feel more certain that it is right. We rely on inspection or checking to make decisions about what to do everyday.

Inspection is not restricted to the manufacturing and quality arenas. Sales people are inspectors, too. You inspect proposals for errors. You study your commission sheet and paycheck for errors. Inspection can provide data about problems or defects, but this has little or no value unless you use it as information for improvement.

Unfortunately, relying on inspection as a means to ensure quality is costly and ineffective.

Why Inspection Doesn't Work

To illustrate the fallacy of relying on inspection to eliminate defects, try this exercise. The paragraph below represents a batch of product that you are planning to ship. The letter "O" represents a defect. Inspect the paragraph below. Identify and count the "O's" just once. Give yourself 90 seconds.

MARKETING: IN ECONOMICS, THAT PART OF THE PROCESS OF PRODUCTION AND EXCHANGE THAT IS CONCERNED WITH THE FLOW OF GOODS AND SERVICES FROM PRODUCER TO CONSUMER. IN POPULAR USAGE IT IS DEFINED AS THE DISTRIBUTION AND SALE OF GOODS, DISTRIBUTION BEING UNDERSTOOD IN A BROADER SENSE THAN THE TECHNICAL ECONOMIC ONE. MARKETING INCLUDES THE ACTIVITIES OF ALL THOSE ENGAGED IN THE TRANSFER OF GOODS FROM PRODUCER TO CONSUMER—NOT ONLY THOSE WHO BUY AND SELL DIRECTLY, WHOLESALE AND RETAIL, BUT ALSO THOSE WHO DEVELOP, WAREHOUSE, TRANSPORT, INSURE, FINANCE, OR PROMOTE THE PRODUCT, OR OTHERWISE HAVE A HAND IN THE PROCESS OF TRANSFER.

(The Columbia Encyclopedia: Sixth Edition. 2000)

How many "errors" or defects" – O's – did you find? More importantly, as a result of your answer, what decisions do you make about what to do with the batch of your product?

- Is the "product" or "service" good or bad?
- What should you do in response to the number of defects?

One response is to take out the defects and ship the rest to the customer. Another way to answer the question would be to turn to the specifications. Suppose the customer says, "We'll take the batch as long as there are no more than 50 defects". Should you ship it or reject it? You may feel good about your answer.

Ask other people to counts the "O's"

- Did everyone come up with the same number? Probably not. Who's right and what is the true number of defects?

  How would the different answers affect the decision to ship or reject?

- How do you know you found all the defects?
  (By the way, there are actually 56)

These are the questions that organizations around the world try to answer every day. People are asked to check things and determine if they are good or bad—and then decide what to do as a result.

During Conway Management workshops, a similar test typically generates responses ranging from 23 to 44, with only a couple of participants coming up with either the right number or the same number. When asked what should be done, the audience responds "count again!" or "average the numbers!" both of which are actions an organization might take. Would that help?

Inevitably someone shouts out, "Ship it!" and everyone laughs. They laugh because they can relate. Everyone knows defective materials are often shipped to accommodate delivery dates. They also know that inspectors can not guarantee quality. Some respond, "But they are catching some of them". **The problem is that the more defects or errors your process produces, the more likely some of them will escape detection.** You have to ask yourself if trying to catch defects is the way you want to run your business. It certainly makes the sales person's job that much tougher.

So what is the alternative to checking and inspecting as a means to ensure quality? *Use the information from inspection to improve the process.* Track errors and problems and use the information used as feedback to find the root causes of the errors, troubles, problems or defects, to change the work so that future errors are prevented. As long as your processes produce errors, you will need to inspect. But you must use the information from the inspections to change and improve the work to truly get value from the inspection.

"Do you think we catch all the defects?"

During a coaching session at a client company, Conway Management met with project teams and the vice president and general manager. The company's product production for a global customer was at a 60% yield. They had 100% inspection at four points in production. They looked at every single item produced at least 4 times and found reasons to reject 40% of the items. The project team had identified a specific cause of rejects and was proposing a process improvement. This improvement would be costly and would fundamentally change the way the product was made. The VP was not receptive. We asked him how long he thought the customer would put up with 40% defective product. His reply was, "The customer doesn't get any defects, and we have 100% inspection!" The inspectors at the meeting were asked if they caught all the defects. Their response was to laugh and say: "No way! We send out as many as we catch!" The VP, who was not amused, said it was the first he had heard of that, adding: "Why am I paying all of these inspectors?" The answer, of course, is because he didn't know any better. He believed inspection was the best way to eliminate defects going to the customer and he certainly did not appreciate the consequences of a 40% defective rate.

Search For Waste: Identifying & Quantifying Opportunities For Improvement

Key points in this chapter:

- The definition of waste
- Why quantifying the waste is so important
- The four forms of waste
- Five ways to search for waste

When we talk about waste, we use the following definition:

Waste: The difference between the way things are now and the way they could or should be if everything were right. Waste represents opportunity!

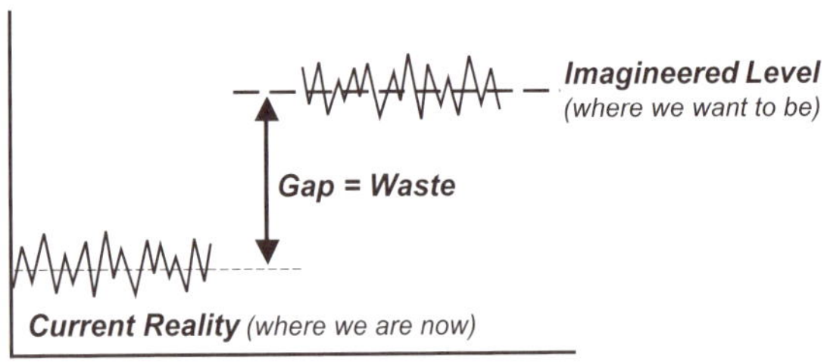

Recognizing the gap or delta between what is and what could be helps you to see how big the opportunity is and creates the desire for change.

During a Conway workshop at a large manufacturing company in Mississippi, it became clear how much animosity there can be toward the concept of waste. One participant stood up in the back of the room and said, "We are not stupid, you know. If there's so much waste, why haven't we been able to see it? Why do I need you to tell me about it?" Others in the room nodded in agreement.

That same week there had been news stories indicating that Pluto was not the outermost planet in our solar system. For years everyone, including the world's leading astronomers, thought they knew what was out there. With the advent of a stronger telescope, however, astronomers could see another, previously undetected, planet. It is not that people are not intelligent, at all. We often need a new lens through which to view a situation or different tools to help see the waste or opportunity.

This is precisely what happens with waste. Using the right tools and knowing what to look for enables you to recognize waste that has been there all along but was not visible. Once people become aware of it, they start looking at it more closely, finding out more about it, and looking for ways to get rid of it.

Once you see the waste, you realize that it is huge and everywhere. Now the important question becomes, of all the opportunities *what do you work on first?* That's when quantifying the waste becomes important. When we talk about quantifying we mean calculating the cost in dollars on an annual basis. We'll talk more about quantifying waste later in this chapter.

Before thinking about how things could or should be, you need a clear picture of the way things are now. This requires an in-depth look at work and work processes. One of the most difficult challenges when working with organizations trying to improve is getting people to study the work. People often want to skip this step. Gaining knowledge about work takes time and discipline. If you do not learn what is going on in the work, you are destined to keep your subjective view of reality, come up with opinions regarding causes of troubles and problems, and develop solutions that may not work.

When people study their own work, they are inevitably surprised by what they find. By using simple charts to paint the picture, the problems become clear and what to do often becomes obvious. Often the group finds that there is no consistent process, that everyone does the work differently. No wonder customers get variation in the product or service!

Identifying Waste Worksheet

Because waste is the difference between where you are now and where you could or should be, you need to answer these two questions if you want to identify waste:

1. Where are you now? (We call this current reality.)
2. Where could you or should you be if everything were right? (We call this the "imagineered state".)

Getting this information will take some time. Here are the kinds of things you need to know to assess where you are today and what the opportunity is:

	Now	Could Be
• What is your market share for each product or service?	_______	_______
• Who are your current customers, by name and volume?	_______	_______
• What are your customer's requirements? Specifically, by customer?	_______	_______
• How well are you meeting your customer's requirements? How could you delight them?	_______	_______
• How many calls are you or your sales people making?	_______	_______
• How do you spend your time on sales calls?	_______	_______
• How long does the sales process take from lead to sale?	_______	_______
• What is your lead-conversion ratio?	_______	_______
• What are the top 3 reasons you lost sales last month?	_______	_______
• How do your prospects learn about your products and services?	_______	_______
• What percentage of your customers/prospects hear about you from TV, radio, and print advertising? Promotions? Direct mail? Website?	_______	_______
• What percentage of prospects call you? What could the percentage be if your marketing were right?	_______	_______
• How many complaints do you receive?	_______	_______
• What is done with complaint information?	_______	_______
• What is your response time to customer inquiries? Complaints?	_______	_______
• How much time do customer service people spend handling complaints? Product/service questions? Warranty questions?	_______	_______
• What are warranty costs? How much do returns, allowances, and discounts cost you?	_______	_______
• How do these problems affect your ability to sell and service customers?	_______	_______

The answers to these questions should be in the form of data, not guesses or opinions. Data enables you to quantify the waste or opportunity, so you'll know if you're dealing with a minor problem, a major problem, or perhaps no problem.

- Once you really know where you are now, picture where you could or should be if everything were right.

Compare where you are now to where you could or should be. That difference represents waste. Assessing current reality and establishing the target or objective enables you to determine the dollar value of the gap.

Four Forms Of Waste

There are four basic forms of waste: waste of material, capital, time and lost opportunity. Thinking about waste in these forms helps to expand your thinking. More important, once waste of these four forms is reduced or eliminated, you can improve the bottom line. For example, if you identify excessive terms or discounts as waste and quantify that waste, you're positioned to make improvements in the sales process, which will improve gross margin dollars. That translates directly to improvement to your bottom line. Being mindful of this bottom-line connection keeps your improvement efforts aligned with your business goals.

Waste of Material – When people think of waste, wasted material is what usually comes to mind. This includes defective products, wrong or defective samples or proposals, wasted energy, inappropriate or excess packaging or marketing materials, and scrap. You might not think this form of waste applies to sales people because scrap is something you see on the factory floor. However, sales people can contribute more subtly to scrap and rework. For example, specifications that are too strict and therefore needlessly expensive are as wasteful as specifications that are too lax and lead to rejection of your product or service. As a translator of customer requirements, your job is to work with others to ensure customer requirements are clear, understood and communicated to the organization and that your organization is capable of meeting the customer requirements. If the requirements are wrong, the specifications and design will be wrong. The result: huge waste in the guise of rework, returns, customer complaints, additional set up times, etc.

Waste of Material

The biggest waste of material comes from not understanding external customer requests or requirements. This can result in over- or under-specifications and engineering designs of products or services that fail to meet customer needs.

List examples of material waste associated with sales and marketing in your organization.

Annual $ value of waste

- Excess or wrong marketing materials
- Duplicate forms or forms where none are needed
- Marketing materials received by wrong or poor prospects
- Waste associated with excess travel or shipments because of mistakes
- ______________________________
- ______________________________
- ______________________________
- ______________________________
- ______________________________
- ______________________________

Waste of Capital — Where is your money tied up that you are not getting full value? This includes excess inventory; overdue receivables; under-utilized offices, plants, and equipment; unfavorable financial agreements; excess capacity.

One major cause of wasted capital in the sales process is reduced prices. When there are problems with quality, a salesperson may offer discounts to improve the price/value relationship with the customer. For what reasons do you "have to" offer discounts? Review your sales orders for the last quarter or month by customer or product list. Compare the prices quoted to what they could have or should have been.

Customer Discount Worksheet

	Customer or Product	List Price	Quoted Price	Annual $ Value of Difference	Reasons for Discount
1.					
2.					
3.					
4.					
5.					
6.					
7.					
8.					
9.					
10.					

Add up the difference. How many more sales do you have to generate to make up the difference? Now review the reasons. If you could eliminate or reduce the reasons, how much easier would it be to meet your sales targets? Who should you work with to make that happen?

For another cause of wasted capital, take a look at receivables. Sales people often spend a lot of time dealing with customers who do not pay promptly or at all. While cash-flow problems can sometimes be a reason an account does not pay on time, more often than not, the reasons have to do with customer dissatisfaction. It could be the product or service was wrong, it was shipped wrong, they received a partial shipment, it was invoiced wrong, or requested documentation was not provided.

What is the current value of your receivables? What should they be? What are the reasons the customer is not paying on time? Are any of the reasons caused by sales? To find out why receivables are not being paid, ask your customers. Most likely, the reasons are due to problems in your company's processes. Other receivable issues include excessively generous terms or delayed invoicing for products or services.

Review the accounts for which you are responsible. Use the receivables worksheet to list accounts whose receivables are beyond terms, and plan to call them to find the reasons.

Receivables Worksheet

Customers beyond terms	$ amount overdue	Reasons for the delay
1. ______	______	______
2. ______	______	______
3. ______	______	______
4. ______	______	______
5. ______	______	______
Total $$ outstanding (waste)	______	

For the organization as a whole, use the following formula to determine the cost of overdue receivables to the organization on an annual basis. For this purpose, consider 30 days receivables "good".

Annual $ sales/12 = 30 days receivables

Actual $ receivables – 30 days receivables = Excess $ receivables

Assume 15% is the cost of capital, additional handling, write downs, etc.

Excess $ receivables X 15% = waste associated with excess receivables

People are often surprised by the magnitude of the opportunity in receivables.

A check sheet is a simple way to track the reasons customers don't pay on time. In Section 3, you'll learn how to convert your check sheet into a Pareto chart that can be used to identify areas for improvement.

Check Sheet

Reasons Customers Don't Pay On Time	Jan	Feb	Mar	Apr	May
Wrong packaging	𝍸 I 𝍸	𝍸 II	𝍸		
Wrong invoice					
Delivery not as specified		𝍸 II			
Inadequate documentation	𝍸				
Other	I		𝍸		

Waste of Time — This is the form of waste with which sales people identify most readily. There are two types of wasted time. One is cycle time (how long it takes to do something from start to finish) and the other is people's time. What time do you spend doing things you shouldn't have to do at all?

Time is a precious resource and wasted time of sales people translates directly to lost gross margin for the organization. Processes that take too long such as developing proposals, obtaining credit approval, and other internal processes, can lead to lost or reduced sales. Time spent on thing such as handling complaints, attending unnecessary meetings, or calling on unqualified leads, reduce the time available for value added sales activities.

Waste From Lost Opportunities - Lost opportunities represent the largest source of waste in any organization. What are all of the things you could or should have done, products or services you could or should have sold, customers you could or should have converted but did not, because of problems with marketing, sales promotion materials, pricing, poor market positioning, lack of communication with the customer, or other reasons?

Basically you want to collect the facts and data about why customers buy or don't buy from you, prioritize them and decide where to work to increase sales and gross margin.

One way to identify this data is with **lost order reports**, which are covered in more detail later in this book. The people who know best why you lose orders are sales people and customers. Work closely with your customers to systematically gather and compile the reasons for lost orders to prioritize what to work on. Lost order reports offer a starting point for improvement because they tell you why current and potential customers for your product or service did not choose you.

Another way to learn about lost opportunities is with the **competitive analysis report**. Who is your competition? Why do customers buy from them and not you? Why do customers buy from you and not your competitors? Again, this information isn't always easy to obtain, but it is critical to increasing sales. One way to tackle this is to regularly designate each marketing or salesperson a competitor to study. Share the competitive analysis with sales and/or management on a regular basis. It shows you the key characteristics you need to improve in order to win more business.

Lost opportunities can also be understood by examining market share. A **market share report** points to the opportunities to increase business. Review market share by sales territory and by customer. How much variation is there and what are the reasons for the variation? Which sales people have a greater share? Why? Do selected products/services have a much greater market share? Why? Why do your customers buy? Why do potential customers not buy? What is your market share in each territory? What percent market share would you like to have? What is the annual dollar value of the difference between actual and desired share? What are the reasons for the differences?

Lost Opportunities

In the spaces below list:

Customers we could have but don't	Reasons	Annual Est. lost revenue
1. ______	______	______
2. ______	______	______
3. ______	______	______
4. ______	______	______
5. ______	______	______

Products or services we could sell but don't	Reasons	Annual Est. lost revenue
1. ______	______	______
2. ______	______	______
3. ______	______	______
4. ______	______	______
5. ______	______	______

Another area of lost opportunity to consider is individual talent. Are you wasting opportunities to share knowledge and expertise? Do you make it in your team's best interest to share know-how and experiences? What are the specific skills and talents possessed by some of the people on your staff that you would like to develop in others? What specific skills do you and others need to acquire to improve your organization's competitiveness?

Wasted Talents

	Talent/Skill	Employee(s) who has it	Lost opportunity associated with not sharing or cultivating talent
1.			
2.			
3.			
4.			
5.			

	Talent/Skill	Employee(s) who needs it	Lost opportunity associated with not sharing or cultivating talent
1.			
2.			
3.			
4.			
5.			

Approaches To Your Search For Waste

In addition to searching by type or form of waste, you can use the following approaches to ferret out waste. The five approaches help you find waste by giving you the opportunity to look at your business from different perspectives. Each perspective highlights different wastes.

Follow The Market, Technology, Competition

This outward looking approach focuses on external factors that influence business success and forces you to compare your organization to others.

Ask questions such as:

- What business(es) are you in?
- Are you in the right business(es)?
- Do you have the right products and services?
- How well do your products and/or services please customers?
- How competitive are your cost and expense structures?
- How do you compare to industry norms or best in class for:
  - Sales per employee?
  - Technologies?
  - Productivity?
- How effectively do you communicate about your products and services? To the right people?
- How responsive are you to customer needs?
- How well do you anticipate customer needs and wants?

Information from this approach can also help uncover major changes to consider in your products and services, technology, and work processes. This could lead you to:

- Change the product or service
- Change the market approach
- Change the technology
- Change the equipment, machines, or computer equipment
- Combine, move, or eliminate processes, plants, offices, departments
- Outsource some of the work or sell a business
- Change suppliers

Follow The Money

This approach helps you to find waste by looking at your budget, Profit and Loss Statement, Balance Sheet and Cash Flow Statement or other financial and operating data. It is the most frequently used way to search for opportunities and often leads to using the other four approaches. The magnitude of the waste is often most apparent with this method. The areas that are in trouble and have high visibility within the organization might be a good place to start looking for waste.

Read the P&L, Balance Sheet, and other financial and operating reports. Ask yourself, where does all the money go? If everything were right, where would it go? Where does all the money come from? How much would there be if everything were right? What would the P&L, Balance Sheet, or Cash Flow Statement look like? The follow the money approach points out where the waste shows up in the financial statements. Keep in mind that financial reports are a reflection of the work processes that produce them. Use the follow the money approach to drive you back into the work to make improvements.

Follow The Systems Or Processes

This approach encourages you to study the systems and process that make up the work. What do they look like? How long do they take? How much do they cost? How efficient and effective are they? What are the critical processes and systems for your department? What systems/processes impact your department? Flowchart the key processes and measure critical cycle times to gather additional process and results data to enable you to identify opportunities for improvement. These techniques are explained in Section 3.

Follow The Time

All work takes time and speed is a source of competitive advantage. Studying time in detail is an important way to find the waste. Time studies or work sampling, are effective ways to find out how time is spent, and are explained in Section 3.

Follow The Ideas Of People

The people doing the work are the experts and are in the best position to tell you about the problems in the work. They often can tell you how to fix them so you need to develop mechanisms to encourage input.

Soliciting employee input is not a new idea. What might be new to you is having a process for managing and implementing the ideas. Two ingredients are needed—a management commitment to actively solicit problems, ideas, and suggestions, and a system to take action on the ideas.

At one time, each Toyota employee on average generated almost 150 ideas per year, and 80+% of these ideas were put into effect quickly. Think of the level of improvement! What could you do to your competitors if you had that level of energy and involvement company-wide?

The Search for Waste

In the space below list all of the examples of waste that you can identify using these search methods. What are the processes associated with this waste? What is the potential dollar value of the waste?

Waste	Process	Annual $ Value of Waste

Human Relations: Encouraging Improvement

Key points in this chapter:

- Creating a culture of continuous improvement
- Simple process checks to eliminate barriers to people making improvements
- An alternative to Management by Objective

> *What a company says or does, the way the sales force is treated, has a major impact on how the organization is perceived by customers, suppliers, and the marketplace. Policies and procedures are a good indication of the organization's value system and fundamental assumptions about what will make sales grow profitably. Our experience shows that organizations will not achieve the level of results possible nor be able to sustain their gains without addressing the systems that drive behavior in the organization. Profit-sharing plans are helpful but an incentive program cannot overcome a human relations system that does not support working this way. No incentive system in the world can overcome poor interpersonal behavior, poor communication and ineffective management systems.*
>
> – *William E. Conway*

Like everyone else, most sales and marketing people are trying to do the best job they can. Unfortunately, current systems and policies can get in the way. These might include current behavior norms, leadership styles or work environment policies and practices that are not consistent with the principles of continuous improvement. An important part of making this management system work is developing a culture that encourages behavior supportive of the goals and objectives of the department and organization.

There are two parts to any human relations system: personal behavior and organizational. The personal side is all about how people treat others. The organizational side is how the people are treated by the organization and the formal and informal practices that encourage people to behave in given ways.

What is considered to be acceptable behavior in the organization? How does your behavior encourage or discourage cooperation and continuous improvement?

Human Relations System

Personal Aspects	**Organizational Aspects**
• Acceptable behavior norms • All interactions - all directions • Create trust	• Rewards • Recognition • Advancement • Education

Let's look at some examples of inconsistencies between current systems and a culture that promotes improvement. Most sales people are not paid to study and improve processes, nor to collaborate with others for the good of the organization. Instead, they are measured and incentivized on their ability to meet sales goals often at the expense of other parts of the organization. Conversely, many sales and marketing people have been unable to meet their goals due to things outside their control including poor territory, lack of marketing, inferior product or service, or poor management. Most of the time the current system, equipment, methods, rules, or policies are causing the problems. Unless management addresses these barriers, they send conflicting messages about improvement.

The objective of examining your current people processes is to identify and remove barriers to continuous improvement by establishing policies and practices that:

- Make it in people's interest to identify, quantify and eliminate waste through process improvement
- Eliminate the waste of human talent
- Eliminate conflicting goals for the individual and the organization
- Create ownership for managing areas of responsibility

Process checks you can use to determine whether a policy or procedure is in danger of becoming a barrier to continuous improvement include:

1. Are people treated as you would want to be treated if you were in their position?
2. Is the policy or procedure consistent with the principle that 90 percent of the time or more a problem is the fault of a process or system and not the individual?
3. Does it make it in people's interest to share information and knowledge about how to do things?

4. Does it make it in people's interest to work collaboratively to eliminate waste?
5. Does it eliminate caps on the quality and productivity of work?

Individual goals and incentives have a long a tradition in sales, and are a powerful motivator. However, you should also recognize where they become obstacles to optimizing sales and organizational performance.

Sub-Optimizing Sales Through Compensation Systems

Many practices encourage sales people to keep information to themselves, rather than make it in people's interest to share what they know so the whole system can take advantage of their knowledge. Sales incentives and bonuses can cause sub-optimization if they encourage divisiveness or misalignment of what's good for the individual vs. the department, division, or entire organization.

The following story demonstrates this point.

"When I first started working in sales, I was working for a resort hotel. I sold to the association market. My sales goal was to sell 1,000 room nights per month. If I did not do that, I would not make bonus. After a few months, I could easily sell 1,000 room nights per month. No matter how many rooms I sold, however, I just turned in the 1,000. I saved the extra 'for a rainy day.' I was getting married and going on a honeymoon, and I knew I would need these extra room nights for the months when my personal life might get in the way of my selling. Luckily that did not happen and my rainy day fund grew. After about 18 months I was promoted. During my last month as a salesperson for the hotel, I handed in almost 6,000 room nights sold! I honestly did not think I was doing anything wrong by hoarding my sales. But imagine management's surprise when they realized their inventory status. How might the knowledge of how much inventory was pre-sold have influenced the marketing strategies and price-setting they had just completed? Unfortunately, the compensation system and the HR system encouraged people to hoard information that could help the organization make important decisions."

What are some things that you do as sales, marketing and customer service that allow you to meet your departmental goals that may cause waste for the other parts of the organization? For example, promising delivery dates that require expediting, or taking an order with incomplete information, causing problems down the line. What are some things that other parts of the organization do that prevent you from reaching your sales goals? Most of these problems can only be solved through working together, process improvement and understanding the true cost to the organization.

The Problem With Sales Goals

The problem with sales goals is that they often drive sales people to meet their target at a significant cost to the rest of the organization. Simply focusing on volume or revenue may result in production, capacity, and delivery problems.

Incentives need to be designed to reflect how the business operates and how to improve it. They should encourage people to share knowledge between departments and reinforce that they must work together to maximize profit.

This doesn't mean that goals are bad. They are an important part of business planning. What is important is that you devise an effective method to achieve those goals that creates a win-win-win situation for the individual, the organization and the customer.

Eliminating Artificial Caps

Dr. Deming listed personal performance reviews, and Management by Objective (MBO) as deadly diseases in organizations because they frequently discourage, more than encourage, improvement.

Some of the dangers of personal performance appraisal are:

- they are too subjective
- they are not done in a timely fashion
- they focus on the person, not the process

Some of the dangers of MBO are:

- It establishes artificial caps on quality and productivity.
- It encourages mediocrity by rewarding people for aiming low.

Our position is not that these are inherently bad. In fact, people work best when they have clear goals and objectives and timely performance feedback. The problem lies with the way these systems are typically used. Personal performance appraisals should be used to help, coach, enable people to do a better job and must take into consideration the effect of the process on one's ability to perform. MBO should to be used to encourage and reward responsible risk-taking and stretch goals, rather than rewarding those who play it safe to be assured of receiving their raise or bonus.

Management by Improving the Process is an alternative to traditional MBO that enables and rewards continuous improvement by focusing not on only on immediate results but also on improving the work processes that allow those results to be sustained. It fosters system improvement and organizational learning.

The goal of any human relations system is to enlist the brains, time, energy and talent of the people to help the organization. Human relations policies and procedures should be designed thoughtfully, keeping in mind the potential for unintended consequences.

Principles Of Work: Understanding & Improving Work

Key points in this chapter:

- Basic concepts and tools for studying work
- The power of the value added work concept
- Why you must study the work to improve it

Improvement is all about studying and changing work. This chapter will provide you with some basic concepts and tools that will help you see your work from some new perspectives so that you can identify and take advantage of opportunities for change and improvement.

Two important points about work improvement:

1. All work can be improved. No matter how perfect it currently appears to be, or how many times it has been improved in the past, there are always ways to dramatically it improve again.

2. The primary focus of improvement must be on the process, not the people. Keep in mind that 90% or more of the problems are the result of flawed systems and processes and not individuals. Most people are trying to do the best they can within the current system. Engaging the people who do the work to study and improve work processes is the best way to improve the results.

Focus On The Work That Produces The Results

Because of the nature of business and the pressure for financial performance, people generally place a strong emphasis on results. So they review the results at the end of every week, month or quarter. When the results are good, they pat themselves on the back; when results fall short, they search for reasons to explain the poor performance. Unfortunately, this postmortem review is used more frequently to justify the good or poor results when using it to learn about and improve the process would be far more useful. Myron Tribus likened using measures in this way to "trying to drive a car by watching the yellow line in the rear-view mirror". The view only tells you where you have been. You need a way to use your measures to predict what your current marketing, prospecting, sales and customer service systems are likely to produce in the future.

Focusing on results is necessary but insufficient for success. At the end of the day, the organization succeeds or fails based on the results. You either generated enough sales to meet the budget and stay in business or you did not. You either got the costs right to be able to make a profit or not. But understanding the underlying causes and using that knowledge to improve work processes that produced those results is where the true benefit of analysis lies.

Some important questions you are trying to answer in studying work are:

1. Are you working on the right things?

2. Are you doing the work the best possible way? How effective is the work in meeting its desired objective and the customers' requirements? How efficient is the work in terms of time and cost?

The Concepts Of Work, Pace & Time

The concept of work requires that you look at the work that you do in great detail. This allows you to get a good picture of current reality. Some questions to ask include:

- What are the tasks or activities you do to accomplish your work? What are the steps it takes to do them?
- What is the purpose of each step or activity?
- Who is involved in doing the work?
- Who is the customer of the work? What are their needs and requirements?
- What materials, equipment, procedures, information are used to do the work?
- What measures are used to track the performance of the work? How are the measures used?
- What is the cost at each step? What is the total cost to do the work?

The concept of pace asks you to look at the rate at which work is done. 100% pace is defined as the rate at which a task or activity can be done over a sustained period of time without causing undue fatigue or errors. Typical examples of 100% pace include walking three miles per hour or dealing a deck of cards into four neat piles in thirty seconds. Both of these examples demonstrate that, for the vast majority of people, 100% pace is very reasonable. Virtually any task can be studied to determine appropriate pace.

Pace is important for competitive purposes because if you are operating at 90% pace while your competition is operating at 100%, you are already at a disadvantage. Unfortunately, people often think that improving pace is just about working faster. Improving pace involves educating about what 100% pace is, then improving the work to remove obstacles that prevent people from operating at that level.

Some questions to ask about pace are:

- How many of a given activity do people do in a unit of time? For example, how many sales calls can a salesperson make in a month? How many proposals can be produced per day? How many accounts can a salesperson call in a week? How many orders can be entered into the system per day?
- What is the variation in the pace from person to person, department to department, territory to territory?
- How can you determine what 100% pace is for each activity you do and improve the work so that you are operating at that rate as a minimum?

The concept of time asks you to look at how long anything takes. Time is a huge source of competitive advantage! And time is a function of the steps in the work, their complexity and the handoffs and delays in the process. Some questions to ask include:

- How long does each step take?
- How long is the cycle time for the entire process?
- How much time is spent in delays or waiting?
- How much time is spent doing inspection?
- How much time is spent in rework or redundant work?

Each of these concepts provides you with aspects of the work to evaluate. Use them when you are doing analyzing a flow chart or doing more detailed analysis of the work such as work simplification or methods analysis which you can learn more about on in Section 3. Once you have studied what the work looks like now, you can ask the imagineering questions to determine what it could or should look like if everything were right. Then you are ready to find out the reasons for the differences and make the improvements.

Challenging The Value Of Work

The value added work concept provides a method of evaluating the nature and value of work. Most work starts out simply enough, but over time steps are added that do not necessarily increase value or meet the needs of the external customer. Many steps may have been added as "Band-Aids" for a real or perceived problem, but they don't eliminate the root cause of problems. The net effect is to add complexity.

If you have a customer-facing job, you may think that all of your work is value added. Unfortunately, that is not the case. The more you delve into the work, the easier it is to see how much of what you do adds no value in the customer's mind at all.

When you study the work, ask which steps:

- Would your customer want to pay for if everything were right?
- Could be eliminated without affecting the outcome?
- Add time, cost, complexity, or opportunity for error?
- Could be changed, rearranged, combined or simplified?

Asking these questions continuously encourages you to evaluate work activities in light of customer needs or requirements and make changes that allow you to spend more time on activities that add dollars to the bottom line.

Categories Of Work

Work can be divided into the following categories:

Value Added Work: This is work that adds worth or value to a product or service from the external customer's viewpoint. Matching customer needs to products or services, presenting proposals, and getting orders may be examples of value added sales activities.

Necessary Work: This consists of tasks that need to be performed to keep the organization operating but have no value to the external customer. For example, completing expense reports, traveling to meetings, doing credit checks.

Unnecessary Work: There are two types of unnecessary work.

Rework—Work that is required only because something was not done properly the first time or work that you must do more than once because your system or process requires it. Rework often is 30% to 50% of total work and is a major cause of waste. Any time you handle a customer complaint, have to interface with operations to explain a misinterpreted customer requirement, enter an order into the system that was faxed or e-mailed to the order department, or expedite an order because of a problem, you are doing rework.

Other Unnecessary Work—All tasks or steps in a process that do not add value and are not necessary for the operation of the organization, such as reports no one needs and/or reads. An example is multiple or redundant approvals or inspections.

Not Working: This includes all the time for which you are paid but no work is being performed. There are two types:

Authorized—holidays, vacations, illness, personal time off, breaks

Unauthorized—waiting time, idle time, downtime, delay time. Waiting time includes: waiting for meetings to start, a customer appointment, price approvals, phone calls, proposals to be typed.

Knowing what category the work falls into helps you decide what the appropriate action is for improvement. Use the following table as a guide.

Category Of Work	Appropriate Action
Value Added Work	• Improve the work • Increase the percentage of time spent doing it
Necessary Work	• Streamline: reduce the time and cost • Improve the quality • Look for ways to eliminate
Rework	• Find the causes and reduce or eliminate
Other Unnecessary Work	• Eliminate
Authorized Not Working	• No action required
Unauthorized Not Working	• Find the causes and reduce or eliminate

Categorizing Your Work Activities

Think back to your last two days of work. In the spaces below list as many of your specific activities as possible. Try to be specific. For example, on a sales call, what was discussed? During meetings, what topics were covered? Then, using the definitions of the various types of work, categorize the work. Try to be objective.

	Activity	Topic/Task	Work Category
1.	________	________	________
2.	________	________	________
3.	________	________	________
4.	________	________	________
5.	________	________	________
6.	________	________	________
7.	________	________	________
8.	________	________	________
9.	________	________	________
10.	________	________	________

The purpose of this exercise is to help you think about your work and how you spend your time.

- Are you working on the most important things?
- How much time do you spend on the things that really matter?
- How much time is taken up by other activities?
- If you made some changes in your work, how much more value added work could you do?
- Whose help do you need to make improvements that are not entirely within your control?

If you were to make a complete list of the activities you do for work and categorize them, you might be curious to find out the percentage of your time you actually spend doing any given activity. A work sampling study is a simple way to collect data on how you spend your time and provides the basis for deciding where to go to work for improvement. You can find out more about work sampling in Section 3.

Imagineering: The Most Powerful Tool

Key points in this chapter:

- The definition of imagineering
- Why imagineering is so important
- How to use imagineering to identify opportunities and to make dramatic improvements
- Key elements of successful imagineering

Imagineering is visualizing the way things would be if everything were right: no troubles, no problems, no errors, no complexities, no waste of any kind. Some of the benefits of imagineering are:

- It helps you or imagine what is possible, whether the change involves making an innovative breakthrough or improving a process.
- It drives you to use a new scale when making comparisons. Instead of comparing current performance to last week, last month, the budget or the competition, it asks you to compare to perfection.
- It helps you to create the gap, which allows you to quantify opportunities.
- Imagineering allows you to look at the work of sales from a new perspective.

Alcoa Corporation coined the word imagineering many years ago, defining it as "letting your imagination soar and then engineering your plan down to earth". The key to making improvements through this method, *is a clear understanding of current reality*. First of all, by understanding the current state, you may find that only minor changes, not a whole new process, are required to accomplish the desired result. Also, when people have a clear picture of the current situation, including what works well and what does not work well, they have far more ideas about the perfect situation. They also have a good starting point for engineering their vision back to earth. Think about it: If you were lost and called someone to ask for directions, the first question would be: "Where are you now?" The more detail you have on your current position, the clearer the route to your destination.

Imagineering can be a helpful tool to develop the perfect sales promotion or set sales goals, budgets, or targets. For the sales promotion process, you might study the current process, including the steps, the time, the cost and the results as well as the problems in the current process. Then, shifting gears from the analytical to the creative side of your brain, imagine the perfect process, quantify the value of closing the gap, and develop possible ways to bridge the gap.

To develop an accurate picture of current reality, learn about the work through work sampling, surveys, flow charts, and the use of the simple charting techniques covered in Section 3. The more you know about the work and understand about the errors, complexities, waste, and bottlenecks, the more effectively you can imagineer the perfect process.

Imagineering should be pervasive and ongoing. Use it regularly to open people's minds to opportunities for improvement. Even people who are not naturally creative can make exciting contributions. By studying the facts and using variation, people can see why the process is not perfect. Knowing those "whys" in detail will make it obvious what must be done to eliminate the troubles and problems.

Key elements in successful imagineering:

- When looking at major processes, imagineering is most effective in groups. Different people bring different perspectives and ideas to the table.
- Enlist the help of those involved in the process to be studied. A group of 4 to 12 people is usually best.
- Maintain an open atmosphere, encouraging and exploring "wild" ideas. Concentrate on perfection, explore the potential for improvement, and avoid shooting down ideas by saying "That won't work" or "We can't do that". Entertain the possibilities!
- Gather facts and data. Imagineering is a flight into the future, but it should be grounded in a thorough understanding of your present position.

As Henry David Thoreau said: "People seldom hit what they do not aim at". If you don't aim for perfection, you will likely never even come close to it. Imagineering gives you limitless opportunities for growth and improvement.

Section 2

Process Improvement Methodology

Overview

Key points in this chapter:

- Why process improvement is the key to continuous improvement
- The methodology for process improvement
- Using the methodology to increase sales

Now that you have an understanding of the principles and concepts of continuous improvement, the next step is putting this knowledge to work.

Because all work is part of a process, it's only logical that in order to improve work you must improve the process. Remember: *More than 90 percent of the time, problems are caused by waste in a process or system—not by individuals.* Your improvement efforts, therefore, should be focused on the process.

As you now know, sales and marketing, like every area of your business, have many processes that are in great need of improvement. Therefore, the same methodology that improves processes in other areas of your organization can be applied to these departments. And, because process improvement eliminates rework and other forms of wasted time, the net effect is an increase in productivity—as well as an increase in sales.

Why You Need A Methodology

The process improvement methodology provides a systematic approach to attack waste. It provides you with a proven pathway to select the right things to work on, followed by a series of steps to guide you to study, change and improve the work. Too often, improvement efforts provide only band-aid solutions to problems. The methodology encourages you to delve deeply into the work to examine processes, uncover waste, and devise permanent solutions. But even though the emphasis is on a logical sequence, the methodology allows you to be creative about the tools and approach at each step, taking advantage of differing perspectives and skills of those involved. The net result is more robust solutions with broad-based support for implementation.

How The Methodology Increases Sales

You can use the process improvement methodology both to improve processes within the sales department and to increase sales.

Some of the most important processes in need of improvement include the process of selecting target markets, identifying prospects, closing sales, and retaining customers. All of these can be systematically studied and dramatically improved.

The methodology trains you to operate in a continuous learning mode. For every action you take, you are constantly getting the facts, analyzing causes where appropriate, exploring the options, selecting a course of action, measuring the effect and making changes as necessary. Applied to sales, it ensures that when a sale is closed, you continue to monitor the account and do not lose your gains. The next time you are given your sales goals, you won't ask, "How will I ever make these numbers?" Instead, you will study the current situation. Who are your accounts? What are they producing? What is the market? Who is going to your competition and why? How much prospecting is necessary to close the gap between current accounts and where you need to be? You will target specific accounts and set a plan to eliminate the waste of lost opportunity. You will also compile data about what is working, what is not working so well and why, and use that information to make more informed decisions.

The Conway Process Improvement Methodology

The methodology is based on the Deming Cycle:

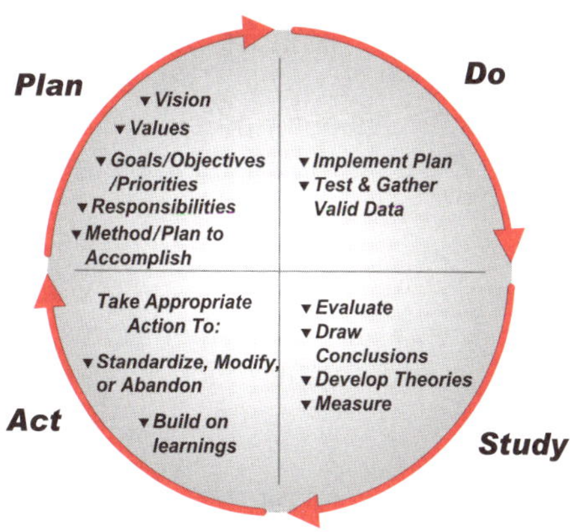

This is a basic model for continuous improvement through continual learning. We have expanded the Deming Cycle to include eight steps that more explicitly describe the critical path to improvement.

These are the eight basic steps:

1. **Search for opportunities; decide what to work on.**
2. **Clearly define the project. Select the improvement team.**
3. **Study the current process/situation.**
4. **Analyze causes; plan the improvement.**
5. **Carry out the improvement plan.**
6. **Study the effect of the changes.**
7. **Standardize the improved process.**
8. **Assess progress and plan for the future.**

While the methodology is basically a linear process, each step is not necessarily a discrete unit. Some overlap may occur and you may loop back to a previous step, depending on what you have learned or how complex the problem is. The following chapters look at each of these steps in detail, and provide sample applications of the methodology.

The same amount of time or detail may not be required for all projects. Once you start looking for waste in processes or lost sales opportunities in your accounts or the market, you'll no doubt discover some relatively easy things to fix. If it's obvious how to fix a problem, do so as quickly as possible. But remember, get the facts before you jump to quick solutions and always use the basic steps of Plan-Do-Study-Act. That is, PLAN the improvement, DO (carry out) the change, STUDY (measure) the effect of the change and ACT on what you have learned. Standardize the improvement if it works well, modify it if it doesn't work as well as you would have liked or abandon it if it doesn't work at all. In other words, make sure that any "quick-fixes" include appropriate process changes, or the problems will recur.

Keys To Success

As you apply the methodology, be single-minded and intolerant of waste. Concentrate on things you can do something about and do not be deterred by those who say things can't be changed or the change won't work. Be willing to run experiments to test your ideas.

Question everything! Check your assumptions and examine your paradigms. Exploring new mental models can lead you to new and profound changes that can transform your business.

Follow the methodology with *discipline*, particularly if you are working with a team of people. Move through each step, taking care to base your decisions on facts and data. Do not skip steps unless they truly do not add value and make sure everyone is on the same step at the same time. Make certain that your investigation leads to *action for measurable improvement*.

Finally, keep the overall purpose in mind: to run a better business by eliminating the waste associated with lost sales and lost gross margin opportunity.

Step 1: Search For Opportunities; Decide What To Work On

Key points in this chapter:

- How collecting data will uncover sales opportunities
- Tools to search and prioritize
- The approach

The objective of Step 1 is to make sure you are working on the right things. In order to do that, you should identify as many examples of waste as possible in the sales area and in areas affecting sales. Invite people from other areas to help you in your search. Then quantify the opportunities in dollars on an annual basis and concentrate your efforts on the vital few areas that will make a significant difference and which you can do something about.

The Importance Of Using Data

You may have ideas about where waste is, and those ideas can help get you started. But rather than speculate, you need to gather information in the form of quantifiable data.

- Data removes opinions from the equation. Sometimes "gut instincts" are right, but not always. And not everyone will have the same opinion. You may think sales are down because of the poor quality of leads. Your colleagues may think they're down because of product quality issues. Your boss may think sales people need to work harder. Data will tell you which, if any, of these reasons are accurate, so you can take appropriate actions.

- Data enables you to be specific, which leads to clear thinking about a problem. For example, instead of saying "I think errors in our proposals are costing us sales," do some digging. Find out how many proposals have errors, the kinds of errors, the correlation between proposals with errors and lost sales, and the amount of sales lost because of proposal errors. Your search may show that proposal errors are not a major factor in losing sales, but that pricing policies or product specs are. Without doing your homework, you can be chasing after the wrong problems.

- Data enables you to determine the extent of a problem, so you can prioritize. If you know that Problem A is costing your department twice as much in lost sales as Problem B, which problem should you tackle first?

- Data helps you communicate the problem clearly to others. If someone comes to you with facts and figures showing that he or she is spending 40 percent of the time handling customer problems, it's going to have a significantly greater impact than if the same person complained that they spend more time being a "pincushion" than selling.

Approach:

- Start by identifying the macro waste in sales. Use the four categories (material, time, capital, and lost opportunity) and the five ways to search (follow the market, competition, technology; follow the money, follow the time, follow the systems and processes, follow the people) to develop a comprehensive list of waste or opportunities.
- Use Lost Order Reports to determine the volume, dollar value and reasons for lost or missed sales.
- Use competitive studies to determine why customers buy from your competition, what they like about your products and services and about doing business with you, as well as what they don't like. Find out what they like and don't like about the competition.
- Compare year-to-year account activity and current sales versus total account potential. Identify the reasons for the differences.
- Use run charts to understand cycles, trends, and the amount of variation around the average. Imagineer what the chart should look like and determine reasons for the gap.
- Use work sampling to find out how much time people spend on tasks or activities that do not add value or prevent you from selling more.
- Ask the customer: what do you want and need?
- Ask sales people: what processes are causing waste? What troubles, problems, errors, and complexities result in fewer sales? What do customers want and need that you do not offer?
- Ask operations what problems/waste the sales department causes because of the way sales does its job.
- Quantify the waste
- Use Pareto thinking to determine what to work on.

Consider quantifying some examples of waste that Paul H. Selden cites in his book, *Sales Process Engineering*:

- Failure of marketing to find and capitalize on existing information
- Insufficient qualification of leads prior to passing to sales
- Blind use of field sales to perform all types of contact
- Absence of presentation impact analysis
- Non uniform and incomplete customer needs analysis
- Pricing policies and systems for complex items
- Labor intensive quote and proposal generation
- Lack of ability to understand and adjust close to true customer motivations
- Order processing errors and waste
- Broken implicit and explicit service commitments

The tools you'll use to search for opportunities are covered in detail in Section 3. One that offers a particularly good starting point for sales is Lost Order Reports. Customer surveys are another important tool. Process flow charts can get you started analyzing a process and check sheets can help you gather data. Run charts give you a picture of variables over time. Work sampling helps you see how much time is spent on work that does not help you increase sales. You'll also need to use imagineering to visualize ideal work, work processes, customer interactions, products and services, etc. This will help you to determine the gap or difference and to quantify opportunities.

Determining What To Work On

Once you've collected the data, you can put it in perspective using Pareto charts. Pareto charts let you rank-order problems or sources of problems. They serve as your guide, pointing to the things you should work on. Study the data. Decide what to work on and in what order.

The biggest waste of all is working on the wrong thing. At least 50% of the Right Way To Manage© is working on the right thing.

Any project or process you intend to work on should have the following characteristics:

- Identifiable waste has been associated with it.
- Clear measures (both process and results) can be established to track progress and learn about the process.
- Its outcome affects external customers directly or indirectly.
- You have clearly defined the start and end point of the process or the part of the process to be studied.
- The project is manageable in size. Break large projects or processes down into bite-sized chunks.
- Changes that are made as a result of your project can be measured.
- What you are measuring cycles daily or weekly so that the effects of changes will be apparent in a short period of time. This is especially important for initial projects but may not apply to later efforts.
- The project is likely to succeed.
- The project has visibility within your organization.
- The process is not currently being studied, changed, or scheduled for study or change in the near future.

Step 2: Clearly Define The Project; Select The Project Team

Key points in this chapter:

- The objectives of Step 2
- Creating a team charter
- Developing effective problem statements
- Developing key performance measures
- The approach

The objective of Step 2 is to create shared understanding of the expectations of the project, to ensure you have the right people on the team and to build commitment for the improvement effort. When defining your project, ***specificity*** is the watchword.

Your objectives are to:

- Clearly communicate the problem and the objective
- Quantify the waste
- Clearly communicate the purpose, benefits, scope, and constraints of the project
- Identify key performance measures
- Select appropriate team members
- Contract with your sponsor for deliverables, timelines, help and support

Team Charter

The Team Charter describes what is being worked on and the expected outcome. It is a written document that defines the project and the contract between management and the team. The charter describes:

- Problem, task, issue, or project
- Objective
- Scope of the project
- Customers and suppliers of the process
- People involved
- Final products

Team Charter Form - Example

Project Definition

Specific item of waste/opportunity: Lost market share of $20MM

Problem statement: Sales of our products are declining while total sales in the market are increasing

Objective: To increase market share or maintain share in growing market

Key business goals which the project supports or impacts: Profitability, brand recognition, growth

Benefits: Company viability, employee satisfaction, return to profitability

Key performance measures:

Overall results measures: Market share, total $ revenue
Process measures: Volume sales calls, % sales to new customers, reasons for lost sales, % sales to active seniors
Quality measures: Cost per sale, gross margin %, image in market place

Recommended Team

Sponsor: CEO

Team Leader: Director of Sales & Marketing

Team Members: 2 Telemarketers, Sales Manager, 7 Store Directors

Facilitator: Outside consultant

Resource Specialist: Vice President Engineering

Project Scope

Background Data: The Z Company invented and has been responsible for the growth and development of an industry with innovative products to assist the elderly. Although competitors have moved into the market, Z has long been considered the industry leader with excellent name recognition and high quality products.

Over the past 5 years, market share and profitability have shrunk dramatically, as competitors have offered good products at a much lower price. Despite new product innovations, we have experienced 2 years of declining sales in a market that is growing rapidly. Current performance trends must be reversed if the company is to remain in business.

Constraints/limits: Piloting changes and implementation require sponsor pre-approval

Resources required: Time for meetings; time for market research

Reporting review/evaluation cycle: Monthly

Target Completion Date: 6 months from start date

Customers: Active seniors, elderly seniors' supermarkets, large chain stores and superstores

Suppliers: Tubing manufacturer, vinyl and leather suppliers, wheel suppliers

Expected deliverable: Improvements implemented

This is an example of a team charter. The initial draft of the charter can be developed in one of two ways. If management at some level has determined that the team is needed, the charter is usually drafted by the project sponsor, the project leader, or a facilitator. This prompts management to clarify their own thinking and level of commitment as well as ensuring that the "big picture" of the work is taken into consideration when deciding what to work on.

For projects that are being suggested to management, the individual or team with the idea may draft the charter. For projects that do not require a team, it is still a good idea to complete a charter before you begin to spend time on the project. This will focus your thinking and clarify what you are tackling.

Many organizations use charters to document all of their improvement projects, whether they involve teams or individuals. This practice helps with tracking, communication and coordination and helps prevent duplication of effort. In any case, the charter should be reviewed, discussed, modified if necessary, then agreed to and signed by the team and the sponsor.

Selecting The Team

The right team is critical to the success of any project. You must have people who know about the work, who have a stake in the outcome, who are interested in participating and who are committed to working together to get results. Every member of the team is essential to its success.

The Team Sponsor

A project sponsor is usually a high-level manager who:

- has agreed to take responsibility for the project
- is committed to its success
- is willing to provide the team with guidance
- has enough authority and influence to provide the team with the necessary resources and to help break down obstacles

A sponsor is absolutely necessary for cross-functional projects and for those requiring resources beyond any team member's scope of authority. Sponsors may not be required for small projects in local areas where the team or individuals are empowered to take action.

Choosing The Right Team Members

When selecting team members, choose individuals who are directly involved in the process being studied or have a vested interest in improvements. These may be people who are customers or suppliers of the process and people who will be affected by the changes. Often, it is useful to have someone from outside the area being studied with a "fresh set of eyes" to ask some insightful questions from a different perspective. In any case, keep the number of team members small, usually no more than four to eight. Use other people as resources on an as-needed basis.

You may need guidance in developing high-performance teams, as well as in creating action plans. See Conway Management's *Team Waste Chasers* for additional information and the Appendix for supplemental products and services.

Selecting The Right Team Leader

The project team leader often plays a critical role in the team's success. As with all other roles, care should be taken to choose the right person for the job. This means someone who:

- has a vested interest in the outcome
- has enough technical knowledge of the issue to ask good questions and critically evaluate information
- has good communication and listening skills to draw out "best thinking" and creativity
- uses appropriate leadership styles to provide direction and create involvement
- can manage both the task and the process
- understands and knows how to use the tools and methodology
- has good organizational skills to manage meetings and project timelines
- has enough organizational knowledge to solicit help and input as required
- is enthusiastic and committed to success

Using A Facilitator

Often it is useful to have someone in the role of facilitator or process observer to partner with the leader in managing interaction in meetings, suggesting and training people on tools and methods as needed, and keeping the team on track with the methodology. Technically, the facilitator is impartial and has no decision-making authority so he/she is free to help the team arrive at an effective solution through the use of a good process. A facilitator can be particularly helpful when the team and/or the leader is inexperienced, when paradigms need challenging or when the discussion promises to be controversial. It is usually beneficial to select a facilitator who is not directly involved with the area being studied, enabling him/her to remain impartial.

Stating The Problem

Developing a good problem statement gets the team off to a great start. The problem statement needs to be clear and succinct. A statement that accurately describes the issue pays dividends by providing laser focus.

Valid problem statements might look like:

- There were X number or percent of missing items on shipments during the month of _____.
- Sales and customer service spent X number of hours researching missing information on orders during the month/ week of _____
- This division lost X dollars of sales due to late deliveries during the month of _______.
- The time from order placed by sales to work scheduled for production averages X days/weeks.
- Salespeople spend an average of x number of hours per week in unproductive meetings.
- Price discounts totaled X number of dollars or X percent of revenue during the period ________.
- X percent of salespeople's time is spent on unqualified leads.

Notice that the problem statements describe what is wrong (errors, time wasted, dollars of lost revenue or gross margin, customer complaints, etc.), how often or how big the problem is and the time frame in question. You can see how this method encourages you to get facts and data as well as to focus on the waste, problem or process rather than the people.

Pitfalls to avoid in stating the problem include:

- "Lack of" or "not enough" statements. These usually describe causes, rather than the result.

  Examples:
  - Lack of training
  - Lack of communication

- "Improving the", "Implementing a", or "Developing a" statements. These usually describe solutions.

  Examples:
  - Improving the training of the sales people
  - Implementing a new proposal-writing process
  - Developing a sales training program

Meeting The Criteria For Good Problem Statements

A clear problem statement meets the following criteria:

1. It states an unwanted effect or result that deviates from the norm. It states *what* is wrong, not *why* it is wrong. Avoid "due to" or "because of" statements, which imply a cause.
2. It is specific and avoids generalities.
3. It is measurable. It says how much, how often, when.

Your problem statement should answer these questions whenever possible:

- How much?
- Who/Where?
- What?
- When?

It should never answer the question "Why?" That is what the team will find out as a result of their investigation!

Key Performance Measures

Key performance measures are indicators of business or process performance. They let you know how well your process is performing and, in the case of an improvement project, they provide you with feedback on whether the actions you are taking are working. Choosing effective measures will focus your efforts and allow you to demonstrate improvement. These measures are usually displayed in the form of run charts, control charts or histograms.

There are three categories of key performance measures:

- **Results measures** show you the overall results of your work. For sales, results measures usually include: Total number and dollars of sales, total gross margin dollars and %, % market share, etc. They tell you the level at which your process is operating, the amount of variation and trends.

- **Process measures** are measures of parts of the process or key variables in a process that contribute to and influence process outcomes. For sales, the number of sales calls per week, response time to customer requests for proposals, lead conversion ratio and percentage of time spent doing value added work are potential process measures. If these move in the right direction, your overall results measures are also likely to move in the right direction. Process measures also help you to isolate where improvements need to be made if you are going to attain your goals.

- **Quality measures** usually tie directly to customer requirements such as % on-time delivery, % order accuracy, etc. Quality measures may also be used to ensure that as you improve something, you don't negatively impact another. For example, you may reduce the turnaround time on requests for proposals but you still need to track percentage of proposals won to ensure that the reduction in cycle time has not adversely impacted the quality of the proposals generated.

When determining the vital few measures for your area or your project, be sure to include both measures of effectiveness and measures of efficiency. Measures of effectiveness relate to how well the work and work processes meet the needs of the customer. Measures of efficiency relate to the time and/or cost of the work or work process.

Operational Definitions

To avoid misunderstandings that can affect your project's outcome, establish operational definitions. These are clear, understandable, agreed-upon meanings for the terms and measures used. For example, if you're planning to eliminate waste that results from discounted prices, you need to define what your project includes. Special package prices? Reduced prices for discontinued items? Volume discounts? Discounts resulting from a history of delivery delays? It may be appropriate to include any or all of these, but everyone involved needs to understand what you are working on.

Creating Problem Statements To Increase Sales

When using the methodology as a tool to generate sales from specific accounts, there should be a separate problem statement for each account.

Here are examples of typical problem statements that could be used to increase sales:

- The _____ market segment has gone from ____ to ____ in sales compared to last year.
- The XYZ account is currently generating $xxx in revenue for our company. XYZ currently purchases from our competitors.
- Territory A is at xx% of market share, which is 30% below all other territories.
- Our scorecard rating from XYZ Company has averaged 3.8 out of 5 for the last three quarters.

By clearly defining the problem or the "presenting symptom", you can systematically use the methodology to delve into the work, uncover the root causes and develop a plan of action to close the gap.

Approach:

- Break down the waste/opportunity into an area or process small enough to get your arms around
- Develop a project charter: Start with the problem statement
- Identify the vital few measurements—i.e., the key performance measures
- Select the team leader and the team
- Come to agreement with the sponsor on the charter
- Check with the appropriate customers of the process: "Is this the right problem?"
- Have we selected the right measures?

Step 3: Study The Current Process/Situation

Key points in this chapter:

- The objectives of Step 3
- How to get a clear picture of current reality
- Tools to use
- The approach

Your objectives at this stage are to get a clear understanding of what the work or work process looks like now, including all the troubles, problems, errors and complexities—and to imagine what the process could or should look like if the problems were eliminated. It is also to identify and prioritize the key issues found as a result of the investigation.

Collecting The Facts About Current Reality

Using your problem statement and key performance measures as a guide, set about to gather information about what the work, work process or situation is right now. Keep in mind that you are studying current reality, not trying to solve the problem at this step.

Start by making a list of the things about which you need facts. Decide what data to collect and what you want to find out as a result. Keep your data collection focused on facts specifically related to the issue at hand. Whenever possible, use data that is currently available.

Organize appropriate data into run charts. These charts allow you to see trends over time and variation from week to week or month to month or year to year. If you have some measures for which there is no historical data, you will need to develop and implement a means for collecting that information. For example, you may have many years of data on dollars of sales per month or number of shipments per month, while there may be no information readily available on turnaround time for proposals or reasons for lost orders.

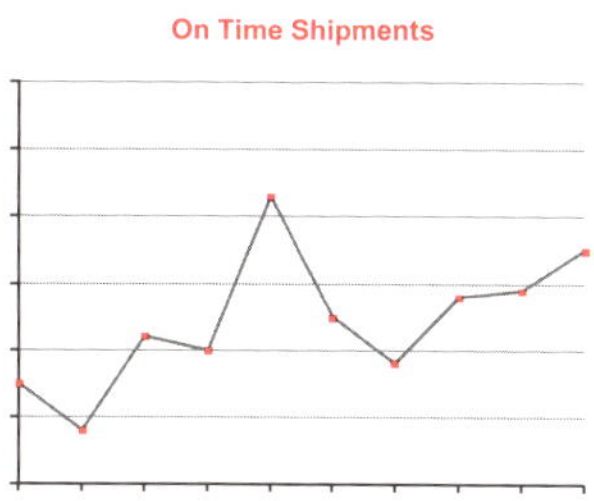

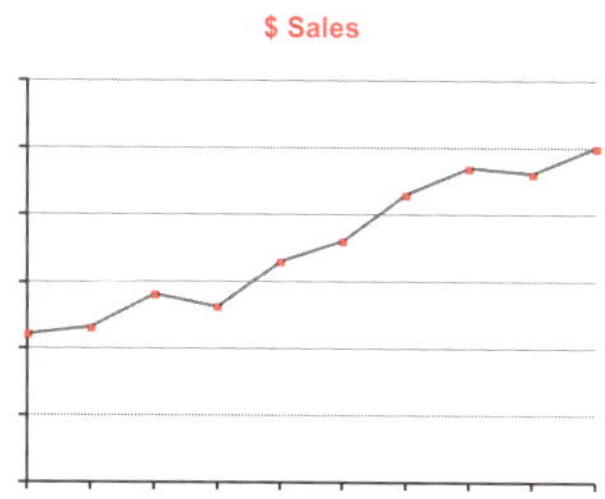

Some of the data may be used to develop Pareto charts. These will help you to see what the biggest contributing factors are, either positive or negative. For example, one Pareto chart might show you which customers bought most of your products or which salespeople or territories had the largest dollars or volume of sales. Another Pareto chart might show you the reasons for customer complaints or reasons for lost orders. In either case, you and your team are trying to gain a common understanding of the facts.

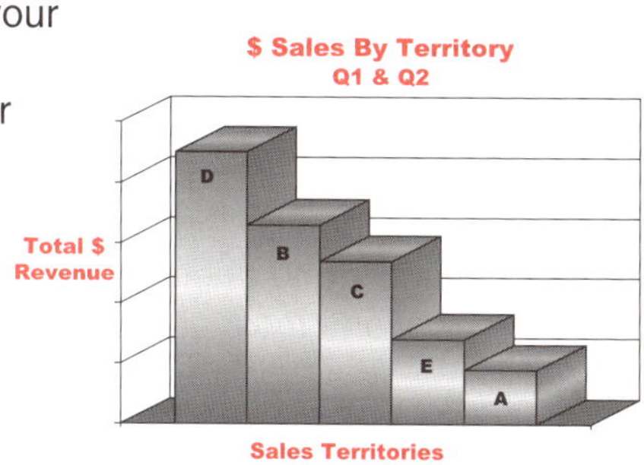

Another essential tool for understanding and improving processes is the process flow chart. Flow charts provide a map of a process. A flow chart of the current process enables people to see what is actually being done, how long it takes for each step in the process, the time between steps, and the total elapsed time. Often, a flow chart of the work exists, but it is not unusual to find that it does not accurately depict the way the work actually happens. To construct an "as-is" flow chart, you must involve the people who do the work. Be sure to include all rework and problems.

Developing an accurate process flow chart is often challenging, especially when everyone who performs the work does it a different way. Even though each person might have a different approach, studying the work will help everyone better understand the work.

After the current process is sorted out, ask the team to imagineer what the process could look like if everything were perfect. This provides a graphic representation of what the work would look like if everything were right and helps you to determine the gap between what is and what could be in terms of time, cost, quality and results.

Once you have the current and imagineered charts, compare them and ask:

- What are the differences?
- What are the reasons for the differences?
- Which steps in the current process are not value-added?
- How much time does each step take? How much time should it take?
- Where do problems occur?
- Why do these problems occur?

Obvious problems that can be fixed immediately should be addressed. Make sure, though, that you are in fact fixing a work process problem and not just alleviating a symptom.

Once you have all of the data and you have thoroughly analyzed it, you should determine the areas to investigate for root causes. Be sure to quantify and prioritize the areas to investigate in order to focus your efforts most productively.

As you proceed through Step 3, review your findings periodically with your sponsor. This way, the sponsor understands the data you are collecting, how you are analyzing it and the conclusions you are drawing. He or she may be able to offer suggestions, assistance or redirection that can improve the team's likelihood of success.

Approach:

- Collect data on key performance measures and display in appropriate charts
- Imagineer the level of performance/variation if everything were right
- Flow chart the current process
- Analyze the current process to determine where the variation, troubles and problems are
- Fix obvious problems
- Flow chart the ideal process. Determine the gap between current reality and the ideal
- Prioritize problems using the Pareto mentality to determine where to focus your efforts
- Identify areas for investigation of root causes

Step 4: Analyze Causes; Plan For Improvement

Key points in this chapter:

- The objectives of Step 4
- How to determine the root causes
- The approach

Your objectives in this step are to determine the root causes of the problem you are investigating—and determine the actions or process changes that would eliminate those causes.

For each issue you selected to further investigate, begin the search for root causes.

Fishbone diagrams, also known as cause and effect diagrams, are good tools to brainstorm all of the possible causes of a problem. Prioritizing the reasons and collecting data using check sheets and surveys to verify the most probable causes points you in the right direction to find to the answer. Once you have determined the most likely causes, you are ready to delve further into the work using a technique known as the "5 Why's".

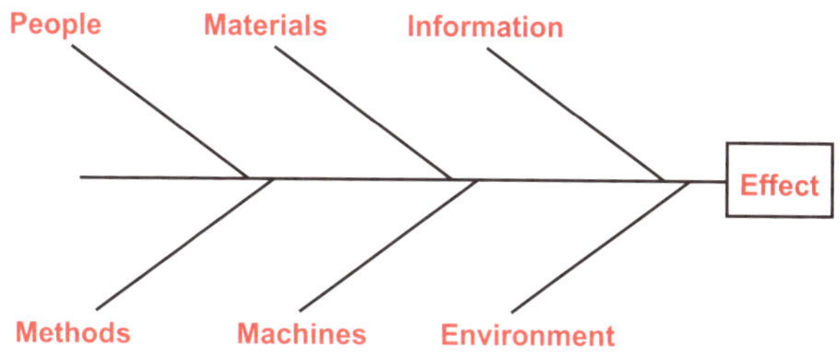

5 Why's Example

We don't have many new customers. Why?
Prospects don't know about us. Why?
We decreased our marketing budget. Why?
We weren't sure that the money was well spent. Why?
We didn't know how prospects/customers heard about us. Why?
We never collected that data on a formal basis and used it to make decisions.

Asking "Why?" several times helps to get to the fundamental or root cause of the problem. Correlation charts or relationship diagrams can help you confirm that you have identified the root cause.

Now you are ready to move into solution-generation. The steps for this are:

- Brainstorm all possible solutions that might address the root causes of the problem
- Evaluate the suggested solutions to determine which will be most effective in solving the problem and fit within the boundaries of your charter. Perform a cost-benefit analysis of the corrective actions. A prioritization matrix is an excellent tool to help you systematically evaluate your options.
- Select the solution(s) to be tried
- Prioritize the necessary process changes, based on your cost-benefit analysis and other important considerations
- Develop a plan to carry out the improvements

The key to success in implementing your changes is to have a detailed plan of action that spells out who will do what and when in order to prepare for and implement the changes. Clarify everyone's role and responsibilities for the implementation. This includes the sponsor, team members, people involved in testing the changes and anyone else who might impact the outcome. The plan should include how you will communicate with everyone and the organization about the change and how it is going, how you will train anyone involved, as well as detailed steps required to make the changes.

Consider the potential types of resistance you are likely to encounter and have a plan for how to prevent or minimize them.

Do some contingency planning as well. No matter how well thought out the solutions may be, there are often unanticipated consequences. For each change to be implemented, brainstorm the things that could go wrong and develop actions to minimize their likelihood or mitigate the effects. You may run into some surprises but you will be better prepared for many of them.

This is an opportune time to review your recommendations with people who will be affected by the changes to get their input, concerns and support. Make any modifications to your proposal then present your recommendations to the sponsor for approval. Once the sponsor has approved the recommendations, be sure to contract with him/her for the support necessary to successfully implement the changes. This often includes help getting resources and using his/her influence to gain support for the changes.

Approach:

- Clearly define problems targeted for root cause analysis
- Identify all the possible causes of each problem and, for the most probable causes, ask "why" five times
- Use additional data analyses to identify and verify root causes
- Determine corrective actions to address the root causes
- Evaluate and select actions to test (try)
- Develop a detailed implementation plan
- Evaluate and develop a plan for dealing with potential resistance and things that could go wrong
- Review recommendations with your sponsor and other stakeholders, as required

Step 5: Carry Out The Improvement Plan

Key points in this chapter:

- The objective of Step 5
- Keys to success
- The approach

The objective of this step is to implement process changes.

Once you've determined an improvement action is appropriate and cost-effective, you're ready to change the process. Whenever possible, try out the changes in a pilot area such as one sales territory or on a trial basis for a defined period of time. This allows you to test out your solutions before implementing them across the board. Make sure to determine how long you need to run your trial in order to have enough data from which to draw good conclusions.

Be sure that your key performance measures are in place so that you can measure the effects of your solutions. Track these measures, also noting anything that might be useful for later analysis.

Carry out your plan with discipline. Make yourself available to people involved in the test or implementation to answer questions, clarify issues or concerns and provide support, encouragement and recognition for their help.

Approach

- Finalize a detailed plan of action
- Make sure everyone understands his/her role and responsibility for implementing the changes
- Make the changes
- Collect data on your key performance measures to determine how well your solutions are working
- Be available to help, coach and thank those involved as they implement the changes
- Let people know how the implementation is going by sharing the key performance measures and other pertinent information

Step 6: Study The Effect Of The Changes

Key points in this chapter:

- The objective of Step 6
- The approach

The objective of this step is to evaluate whether or not the changes you have made have resulted in the desired performance levels and to make changes or modifications as necessary to achieve the goal.

Track the key performance measures identified in Step 2 that will enable you to determine if the change is having the desired effect. Use run charts, control charts, histograms, or compared Pareto charts to monitor this. Use careful analysis to help you draw appropriate conclusions.

If the key performance measures are at the desired level, standardize the process so that everyone does it the same way (see Step 7) and so that you can hold the gains.

If the key performance measures have moved in the right direction but are not where you would like or need them to be, analyze the reasons for the differences, develop modifications or additional corrective actions, and develop and implement a plan for the modifications. Continue to measure to see if the modifications are working. If you're unable to reach the desired level, you need to delve deeper into the process. Return to Step 3 to revisit the process and identify other areas for root cause investigation.

Or, if you think you've missed the mark entirely, abandon the change, go back to the beginning, gather more information and redefine the problem.

Approach:

- Analyze key performance measures to determine the effect of the changes
- If the desired performance level is reached, move to Step 7
- If the desired performance level is not reached, return to Step 3, 4 or Step 1 as appropriate

Step 7: Standardize The Improved Process

Key points in this chapter:

- The objective of Step 7
- The importance of standardizing
- The approach

Your objective now is to "institutionalize" the improved process. Standardizing the improved process permanently removes the root causes and allows you to hold your gains. The process of standardizing also sets a new performance baseline from which to make additional improvements.

For an improvement to be long lasting, everyone affected must be informed of the change, verbally if possible, as well as with written memos or other documents. Make the change as easy to follow as possible. Don't create unnecessary forms or paperwork. Remind people how much the waste was costing the organization and why it is in their interest to implement the improvement.

A Culture Of Standardizing Improvements

When I worked for Marriott I saw process implementation in action. I was the pre-opening Director of Sales for a 350-room hotel in Kansas. I hadn't worked for Marriott before and one thing that I noticed immediately was the complete adherence to procedure. It was such a part of the culture that employees knew SOP's (Standard Operating Procedures) by number! It was not uncommon to hear someone say, "This is covered by SOP #47." As guidelines, the SOP's were simply and clearly written and often included flowcharts that showed pictures of the work. The SOP's were divided into two categories: SOP's and LSOP's. LSOP's were local standard operating procedures. Each property could identify processes that were unique to their property and develop a procedure. This was rarely done by one person and always approved by several people or a committee. A manager would *never* just decide on the fly how something would be done. For example there were few handwritten notes in account files. Sales people carried a Dictaphone recorder with them. They would record call reports, orders, proposals, letters, and client comments into this recorder and a transcriber would transcribe all of this data/information. The information would be sorted and distributed along established distribution channels. This increased the amount of data collected and also reduced the variation. If someone just dropped a note into the file it would not be uncommon for a co-worker to say, "that is not SOP, that note should be distributed throughout the entire process.".

Years after I left the Marriott and was with Conway, I was working with a hotel client that had a copy of the Marriott Sales and Marketing SOP book. They were so thrilled to have the "holy grail". I am sure Marriott was not thrilled to have this volume in the hands of a competitor. But, I don't think that they were too worried either. I said to the client, "You know, John, the magic of this book is not really in the processes themselves. The magic is that Marriott's culture is such that everyone is devoted to following and understanding these procedures. They really have faith in them and believe that it is wrong to work outside of the process. At the same time, what makes it even more powerful is that every employee at every level is watchful for improvements and understands how to communicate improvement ideas so that they are studied and implemented."

To standardize the improvement:

- Prepare a new flow chart, if appropriate, to show what the process now looks like. Post or otherwise distribute copies of this flow chart.
- Revise or develop standard policies and procedures handbooks or instruction sheets to reflect the process changes. This includes formats for continuing to measure process performance.
- Communicate the changes to people who are doing and managing the work.
- Train people in the new method or procedure.
- Continue to monitor the processes to ensure that the changes are maintained. Provide help and coaching where necessary.
- Look for other areas where process improvement changes can be replicated or where the lessons learned can be applied.

Approach:

- Document the new process with flow charts, standard operating procedures and any other useful mechanisms
- Communicate the changes to everyone involved
- Implement training as necessary
- Continue to measure and monitor the process
- Look for other areas where the improvements can be replicated

Step 8: Assess Progress & Plan For The Future

Key points in this chapter:

- The objectives of Step 8
- Recognizing and learning from team efforts
- Determining "next steps" to increase sales
- The approach

The objectives of this step are to evaluate accomplishments, determine future actions and celebrate success!

Step back for a minute and look at your team's progress. Identify specifically what was accomplished and quantify it. Whether you've increased sales, reduced costs, saved time, or improved lead conversions, you're strengthening the bottom line as well as your ability to remain competitive.

At this point you also want to determine next steps. Based on what you now know about the process and its level of performance, are there additional gains to be made that warrant attention? If so, and this is an important item on the organization's Pareto chart of opportunities, charter a team to attack the waste. If not, make sure that the standardization mechanism is in place to ensure that the results do not deteriorate.

Celebrate your team's success and make sure your team's efforts are recognized. Publicize the results. Internal newsletters, company-sponsored closed circuit TV, intranet announcements, presentations to top management and the organization, thank you notes and postings on bulletin boards all offer ways to spread the word about successes. Recognize everyone who contributed to the success of the effort. This reminds people that everyone is in this together and it helps motivate others to participate in improvement efforts.

This is also a good time for you to assess your team's performance and identify key lessons learned for yourselves and for the organization.

Ask:

- What went well?
- What could be improved?
- What could be done differently?
- What training will benefit the team?
- What lessons can you share with other teams that would help improve their performance?
- What lessons and recommendations can you share with the organization that will help other teams be even more successful?

Next Steps

Having completed one project, you're now ready to move on to the next challenge! What is the next area to be attacked that will reduce the waste of lost sales, lost opportunities and lost gross margin?

To determine next steps, review the problems and causes that were identified during your project work but not acted on. Decide whether to pursue them. Do a cost-benefit analysis to justify further action. Or go back to your initial Pareto chart of opportunities and decide what to go after next. In any case, remember that continuous improvement means just that: it goes forever!

Approach:

- Evaluate team progress and overall accomplishments
- Celebrate and recognize the team's efforts
- Review team performance for lessons learned
- Decide what actions to take next

Process Improvement Methodology In Action: Sample Project

This section includes an examples of applying the methodology to sales, with a focus on waste from lost opportunity.

Background
The founder and CEO of Z had basically invented a new industry with an innovative product aimed at assisting the elderly. Over twenty years, the company had grown and prospered as new and better products were added to the line. Naturally, the profitability and potential growth for this market attracted competitors, but Z had an established name, excellent products and was considered the leader in the industry.

But over the past 5 years, Z has watched its market share shrink dramatically as the newcomers to the industry offered good products at a much lower cost. Despite Z's careful and innovative engineering, they had had two years of declines in sales despite the fact that the total market was growing quickly. The declining revenue did not offset the fixed expenditures and profits evaporated. Incredibly, the company that founded the market was at risk of going out of business while its competitors thrived. Z needed to sell more and grow! They decided to tackle the situation with the 8-step process.

Step 1 – Decide what to work on
While there were many areas of improvement to be made, increasing sales volume was clearly the most critical. To quantify the waste, Z calculated that if they had only *maintained* the market share they had the previous year, revenue would have been 30% higher. If they had the market share they'd had 5 years previously, the company would now be close to three times its current size. Assuming they should have been able to maintain market share at current margins, the opportunity cost of the lost profits was approximately $20MM this year. They decided they *had* to work on Sales.

Step 2 – Clearly define the project. Select the improvement team
This step seemed simple enough to everyone, at first – but that proved not to be the case. The way an organization defines the problem often opens some avenues for exploration and closes others.

A mistake in defining the problem will dramatically reduce your chances for success.

At Z, several problems statements were proposed and discussed. The CEO and founder of the company said that the problem was that the sales force didn't understand the products well enough to persuade the customers to buy them. The Director of Manufacturing thought maybe they weren't charging enough for the products – perhaps they had a pricing problem as much as a volume problem. The head of the Z Stores division pointed out that a key competitor offered certain new features that Z's products lacked; but the CEO dismissed those features as unimportant to their customers – many of whom, at least in the early years, he had known personally. The Director of Sales and Marketing believed the problem was that Z's products were simply priced too high — 25-30% more than the competition's comparable products. But the founder pointed out that they were not truly comparable. The cheaper, competing models were not nearly as durable and well engineered and they did not have the name recognition Z's products had. He said that just proved that Sales just did not understand the product merits if they thought they were over-priced. After a moment or two of glum silence, the CFO wondered if the Sales folks were making as many sales calls as they should be. Perhaps the competition employed a more motivated sales force.

Each of these proposed problem statements carried an implicit assumption about the cause of the problem — but none of the assumptions had been tested with facts and data. They agreed on the following problem statement: "Sales of our products are declining while the total sales in the market are increasing."

The management team pulled together a group of key individuals across the different functions, and named the Director of Sales and Marketing as the project leader. The project leader was to develop the charter, determine if additional team members were needed based on the complexity and size of the project, and establish the key performance measures.

They decided that the key results measure would be sales revenue. They discussed establishing market share as the result measure, but that data was available only annually. They decided to measure sales, with the target based on assumptions about maintaining share in a growing market.

They also discussed what their process measurements should be and discovered they could not establish process measurements until they better understood what a successful process in today's environment should look like – and that would necessitate understanding the root causes of their problem.

The team's progress would depend on how quickly and accurately they could factually test and determine the causes.

Step 3 – Study the current process/situation
Z set out to get the basic facts and data about the problem. They made a grid of competitors' products compared with their own: the different features, the different prices, the different sales volumes and sales volume growth rate. They mapped out their three primary sales channels at a fairly high level and labeled these maps with the sales volumes, profit margins, products, and number of sales and administrative people involved in each sales method. They annotated each of the high level maps with information about which competitors use a similar process and what volume they flow through that channel. They gathered feedback from their customers: customer surveys, customer complaints, and letters the customers sometimes sent in to the founder.

Step 4 – Analyze causes. Plan the improvement
Once you have comprehensive current factual information, you can identify causes for declining sales and market share. Step 4 has three components: identify possible causes, test those possible causes with facts; and plan improvement.

First, identify possible causes. Z's management team had begun with a number of ideas about causes, as we saw when they tried to write a problem statement. In fact, several were so convinced of their ideas about the main cause that they were impatient with steps two and three in the problem solving process. They began by listing the causes they believed were contributing. Then they turned to testing and verifying.

Second, test those ideas about possible causes with facts. Verify the most significant causes with data.

As is often the case, Z found some significant new information when they set out to test some of their pre-existing ideas about the causes of the sales decline. Here are some of the things they did.

When they looked at the grid comparing their products to the competition, they noticed a substantial price difference for no real differences in features. Were the customers willing to pay a 25%-30% premium (about $900) for the highly respected brand name? Some on the management team believed they would, and others strongly disagreed. The Finance person pointed out that they were barely breaking even at the current prices – how could they possibly match the competition's prices?

The team figured out that they would have to increase the volume by at least 10% to compensate for a 20% cut in price – or they would worsen their already poor financial picture. The Sales and Marketing Director believed they could do it and suggested they conduct the test in a sales area that was not seeing much action anyway, so the costs would be small if his hunch was wrong and volume did not rise substantially. The CEO was sure he would fail, but agreed that this was a low risk test that could bring closure to this perennial Sales complaint. So they conducted the experiment. They cut the price on their low-end model to just slightly above the competition's in the low volume, low risk sales area.

Within eight weeks sales volume had increased 50% on the model when it was priced close to the competition – simultaneously increasing profit and market share. The team concluded that the market was willing to pay a premium for the established name, but that the premium had to be much smaller – at least in some areas with some products. They concluded that their pricing was perhaps a major contributor to the loss of market share to the competition.

They also considered how to test whether product design was a factor. The VP of Engineering was interested in the suggestion, made by the director of the Z Stores, that some of the competitor's new enhanced portability features were contributing to the loss of market share. To evaluate the popularity of the portability features, they would need to talk to the customers – preferably to those that were buying from the competition. But with cash running short, the CEO balked at conducting a market survey, and instead suggested that the customer service department test customer sentiment about features when they are taking orders. These results were consistent with the CEO's opinion that the product designs were not part of the problem, and the team moved on.

Some time later, the company did commission a fuller market survey that produced more complete and somewhat surprising information. The survey confirmed that the more elderly and infirm customers – which had always been a large portion of the Z's customer base – placed very little value on portability features. However, the broader survey showed that the newest and fastest growing segment of the total market was the "active seniors" and that this segment placed a high value on product portability.

In retrospect, the team saw that while the lack of portability features did not contribute greatly to declining *revenue*, it was a major factor in the declining *market share*. If they had examined the facts and data more rigorously up front they may have saved themselves a costly delay in addressing the problem.

The team also tested the possibility that the sales force was not making effective use of their time. To get the facts they enlisted the sales folks, mostly telemarketers, to study their work using the work time analyzers. They learned that the telemarketers spent a surprisingly low portion of their time actually talking to customers, and a surprisingly high portion of their time talking on the telephone within the company. The team concluded that this could indeed be a major cause of declining sales.

It was suggested that the telemarketers be given new measurements and incentives to increase their customer call rate (the implied root cause being lack of motivation or incentive). But before they acted on this, they decided to collect more data about whom the reps were calling internally and why. They designed a data collection sheet, gathered the data and produced a pareto of the reasons for internal calls. These indicated specific process problems that resulted in the need for frequent intervention by the reps so as to get an order into the pipeline. They were starting to get closer to root causes.

Z's analysis indicated two major causes to address to increase sales and market share: more competitive pricing and more productive sales reps. The latter was one of those quick breakthroughs that often happens when an organization gathers data and sees a problem in a new way. Once the pareto of causes for internal calls was analyzed, they moved quickly through Step 5 – Carry out the Improvement Plan. The underlying causes were quickly countered with better forms and screens, reduced variability, and anticipating frequently asked questions. In very short order, the sales reps were able to more than double the time spent actually talking to the customers.

The results of the pricing experiment challenged some deeply held beliefs and required more careful analysis as well as action plans. Z's management team raised and tested additional questions. Would the results in this one area apply to other sales areas as well? Would it follow that some of the mid range and high end products were overpriced as well? If they changed their pricing strategy, they wouldn't get the margins they had always believed necessary to profitability. Should they change their margin targets or change their cost structure? Could they manufacture as productively as their competition? Would it be wiser to work on reducing costs before making any changes to prices?

In the past, they would have debated these questions and then come around to agreeing with the CEO's opinion. But they were committed to following the 8-step process and decided to resolve the questions by going in search of more facts. In doing so, they developed a deeper and more comprehensive understanding of what today's customers wanted and what they would pay for, and were able to design and implement an improved pricing process.

Step 5 – Carry out the improvement plan

For Z, instead of changing a price or two, they worked to change their *pricing process.* The process had been a cost driven process. A price-range would be set in the specification stage based on what design engineering believed it would sell for — then cost targets would be derived. But the final price would be driven from the final cost plus a standard mark-up. Costs would always creep up so that the final price was at or above the upper end of the price range. The process consistently resulted in uncompetitive prices. They needed to do more than change a few prices, they need to change to a market-driven pricing process.

Step 6 – Study the effect of the changes

Z studied their process changes and discovered that indeed the sales reps were making significantly more customer contacts and this resulted in increased sales. They were able to establish on-going process measurements around the number of customer contacts per day so as to identify and address any declines in contacts (which would happen as new problems crept into the process) before the problems showed up in declining sales.

They continued to study and improve their pricing process. They incorporated periodic process checks to ensure that the new pricing process was working effectively to maximize both profits and market share. And they began an aggressive effort to design more efficient processes throughout the company in order to succeed with more competitive prices.

Step 7 – Standardize the improved process
Z found that while the initial results with the sales rep's time were great, after a couple of weeks the sales reps were making more internal calls. They needed to continue to work and improve the process until all the bugs were worked out and everybody was fully trained and experienced with working the new way.

Once they had "finished" the new pricing policy, they had some rough going before it was standardized. The challenge was to stick with the new process rigorously enough to measure it, but remain open to proposals for additional improvements. If individuals implemented their own improvements without sharing and gaining agreement with them, the process itself would have too much variation to be measured and managed.

Step 8 – Assess progress and plan for the future
In this example, Z had the goal of increasing the sales and market share. When they assessed their progress, they found that while the sales rep productivity and the pricing improvements led to increased sales and slowed the loss of market share, these improvements were not sufficient to reverse the declining market share. The management team had to go back and study the situation further, gathering more information to better understand other possible causes.

In their next attempt, they commissioned a thorough customer survey and learned that their products and services were most popular with the elderly seniors — the slowest growing segment of the market. This was the market segment that the CEO had first addressed, investing personal time and energy in learning all about his customers needs. He never forgot what he had learned first hand and these customers were happy and loyal. But as the market expanded to include active seniors, Z's products and services continued to be geared toward the elderly – missing a growth opportunity. What is more, as the active seniors became more elderly, they remained loyal to the competitors whose products had so pleased them in their more active years.

This information helped Z to change course from a path that was leading them out of the business. They realized that they had to actively pursue the facts about the active senior's preferences and make the necessary changes to their products and services.

Section 3

Tools For Improvement

Getting Started With Data-Gathering

One of the first questions you need to ask yourself is: "What data should I collect?" The short answer is data that will help you identify: what to work on, the size of the problem or opportunity, the reasons for troubles and problems, how to make improvements, how well solutions are working and how to sustain improvements.

Starting Point: Current Reality

For sales, current reality is your account and market situation. To really understand where you are now, you need an account coverage program that measures what your current accounts can be expected to generate in sales. Just knowing last year's sales is not enough.

The basic knowledge you should have for each of your products or services includes specific information about:

- the product or service
- your accounts
- your market and its segments
- your competition

Basic knowledge of your current account situation will be the basis for forecasting sales, staffing, and understanding how much marketing will be needed to generate new accounts to fill the gap between what you have and what you need to make your sales goal. The one consistent problem statement for all sales people is:

"Currently my account base produces $_____. This is $____ below expectations."

The basic information every sales and marketing organization should have includes:

- **Market share**
  Data by major product/service line for each territory, region or marketing area and each major customer as appropriate. This can be an excellent starting point for identifying what to work on. Where there are major differences among areas, you will want to learn the reasons for the differences.

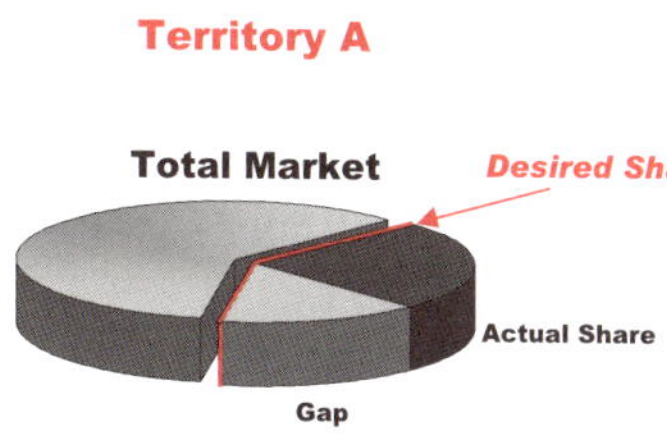

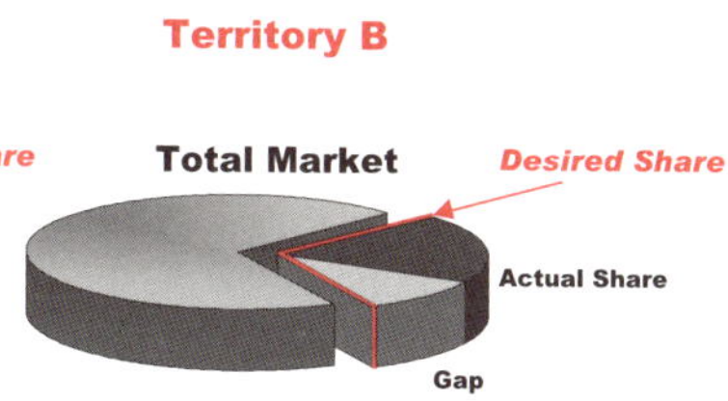

- **Market research**

  If you want to be sure that your product or service will fulfill a need in a way that outdoes your competition, you need to find answers to these key questions: What do customers value? What would they buy? Why? Who is the competition? What do they offer? What about them is more attractive, less attractive or comparable to what we offer?

  In most cases you'll need market research. For example, Philip Kotler, author and economist, wrote in *Marketing Management*, "Coke knows that we put 3.2 ice cubes in a glass, see 69% of its commercials every year, and prefer cans to pop out of vending machines at a temperature of 35 degrees. (Other marketers know that)...we each spend $250 per year on flowers; Arkansas has the lowest consumption of peanut butter in the United States; 51% of all males put their left pants leg on first, whereas 65 percent of women start with the right leg." This is the type of information that can be gathered about prospective and current customers. You may not need this kind of detailed information about your customer but to be competitive you must know more than just the what, where, when and how information about the customers.

  Market research can be both formal, conducted either internally or by an outside firm, or informal where sales, marketing and customer service are constantly gathering and compiling market "intelligence".

  This data, like all data, can be expensive and time consuming to gather so it is important that it be targeted. Market research is as important to the sales and marketing effort as product research is to product development. Well-conducted market research provides valuable insight that can be used as the basis for improvement efforts. Before you embark on any market research effort, be clear about what you want to know. In addition to the information included above, market research should be continuously collected on:

- Buyer responses to new products and product improvements
- Buyer evaluation of advertising, packaging, and brand positioning
- Effect of pricing on buyer purchase intent

- **Reasons for lost sales**
  To grow your sales, you need to know why you lose sales now. Use a systematic way to track lost orders and the reasons for them, then accumulate that information and use it for improvement. Share that information with other parts of the organization that can impact the reasons you lose sales. Here are the results of one organization's lost order reports, with results shown over time on a run chart and the reasons for lost orders shown on a Pareto chart. The client believed (correctly) and verified it with data that if they could improve their delivery performance, they would lose fewer orders.

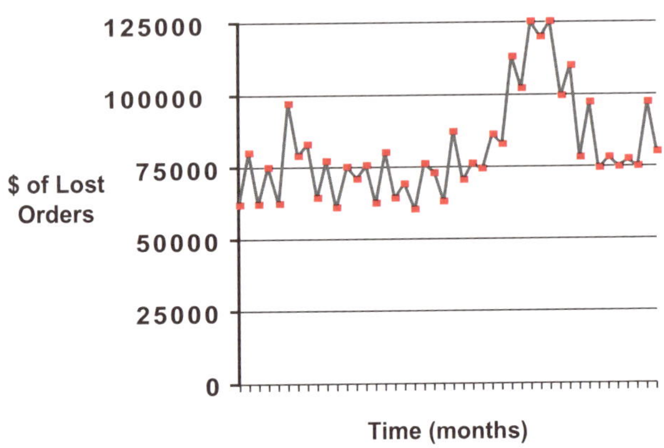

Data can be gathered weekly using a simple check sheet. Break the information down by major product or service and communicate the results using a Pareto chart. Obtaining valid data on reasons for lost sales may be difficult, so be sure to distinguish between hard data (facts) and soft data (opinions).

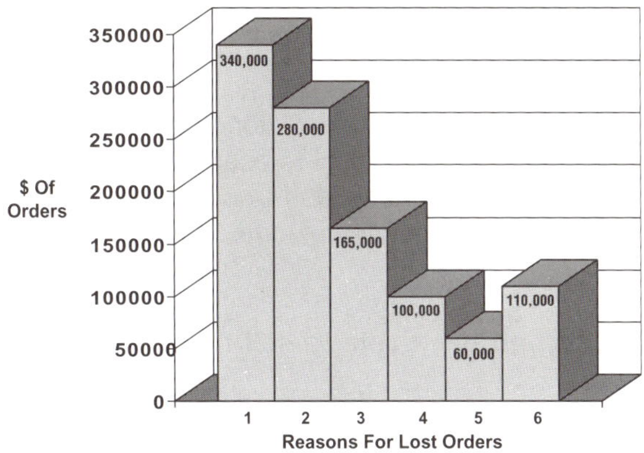

Reasons For Lost Orders

1 *Can't meet delivery date*
2 *Competitor product size can be easily adapted to their machines*
3 *Price too high*
4 *Don't like our documentation*
5 *Didn't know we were in that market*
6 *Other*

Another helpful tool is gap analysis. First you need to know which characteristics/qualities the customer views as important and <u>how important</u>. Then you need to know how your product, service or organization compares to what the customer is looking for. The following chart shows this information.

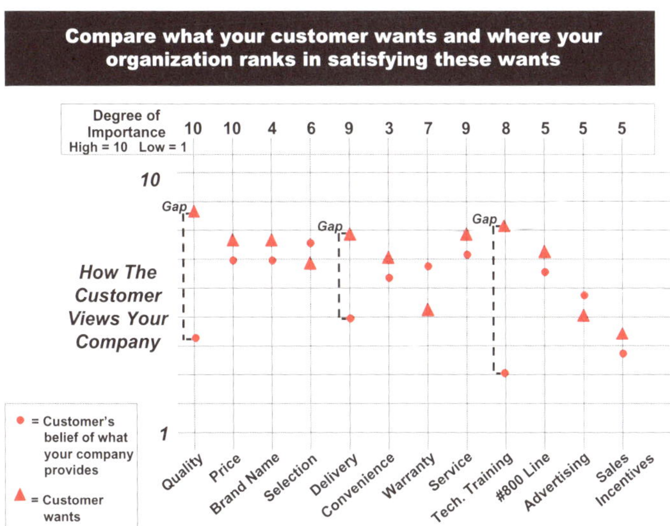

The characteristics with the largest gaps and the highest degree of importance provide real opportunities for improving customer satisfaction.

- **Data on the work**
  A thorough understanding of work processes guides your efforts as you strive to improve the work and increase the percentage of time you spend doing value-added work. Detailed knowledge of the work improves the quality of forecasting, sales plans, lead generation, generating proposals and closing sales, while enhancing your ability to handle customer complaints.

Questions to ask to find the waste to sell more

The following list suggests questions to ask to find the waste to sell more and grow.

1. What are your current service processes? How effective are they in meeting customers' needs? How long does it take you to serve your customers? How much more could you sell if you could improve the effectiveness and reduce the time and cost?

2. What is your current product/service list? How well do your current products/services meet customer needs? Wants? What unmet or emerging needs represent opportunities?

3. How much does your product or service cost to produce/deliver? What could or should it cost if everything were right? (Where are the material, time and capital wasted?)

4. What is your marketing process? How effective is it in creating demand and name recognition?

5. What is your advertising process? How effective is it? Placement? Frequency? Timing? How do you determine if your advertising dollars are well spent?

6. What is your promotional process? How effective are your promotions? What is the return on investment for each promotion?

7. What is your current distribution process? (OEM, Distributor, Dealer, Retail) Are you using the best channels of distribution for your products and services?

8. What is your lead development process? What is the process to qualify leads? Do you have enough qualified leads? The main purpose of advertising, promotion, trade shows and telemarketing is to get leads. How effective are your lead generation processes?

9. What is your current sales process? To current customers? To acquire new accounts? How effective is it in helping you get business? What works well about it? What hinders performance?

10. What is the process for determining optimum pricing? How do you know your pricing delivers optimum revenue and profitability?

11. What are your sales/marketing, customer service policies? Do they meet the needs of your customers? How easy do they make it to do business with you? What are the big opportunities for improvement? Do they meet your organizational needs?

12. Do your current new product/service development processes provide your organization with leading edge products and services? How long does it take to commercialize new products and services? How effective are your product launches in generating new and continuing business?

13. How do you use technology, including the Internet to find prospects? new customers? new marketing ideas? communicate with customers and prospects? How does your use of technology impact sales productivity?

The first part of this section describes the tools that enable you to collect the data you need to start your continuous improvement project work. The second part covers the charting techniques that help you convert your data into useful information.

Data-Gathering Tools

The primary data-gathering tools that will provide the facts you need are check sheets, work sampling, and surveys. Data should be collected by those directly involved in the process being studied. Although people often say they are too busy already to take on the added responsibility of collecting data, their concerns diminish when a simple mechanism is designed for gathering the information and when they receive feedback on what is being learned and what action is being taken. They also need to be assured that the information will be used to understand and improve the work, not to blame them.

Data gathering does not have to be an overwhelming burden. It's often a case of simply tracking information already available to you, such as customer complaints. Even when new data-gathering efforts are required, the payback is well worth the additional time spent.

Data Collection Tips

- Keep your purpose in mind.
- Carefully plan the data collection effort.
- Have clear operational definitions of data to be gathered.
- Collect only relevant information
- Use historical data first.
- Keep your data collection forms simple.
- Determine how much data you will need in order to draw valid conclusions.

Check Sheets

A check sheet is a tool that lists the things you are interested in and enables you to record each occurrence of the item. It does not reveal the sequence in which the data was taken, nor does it qualify the results in any way.

Making A Check Sheet

A check sheet can take several different forms. It uses check marks, "x's," or tick marks to show a count of items falling into various categories.

Here is an example of a check sheet:

Check Sheet
Calls from Customers - August

Reasons For Calls To Chicago Sales Office					
Customer	Product	Pricing	Complaints	Delivery	Total
A	XX		XX	XXX	7
B	X	XXXX			5
C	X	XXX	XXXXX	X	10
D	X	XX	X	XX	6
E	XX		XXX	X	6
Total	**7**	**9**	**11**	**7**	**34**

Questions to ask

- What is the purpose of the check sheet? What are you trying to determine?
- What will you do with the data collected? What chart(s) will you want to convert it to?
- Is the type of check sheet appropriate for the analysis you want?

- Is it simple to use?
- Does the person using it understand the importance of recording all the data and how it will be used?
- Does he or she have a clear operational definition of the data being collected?
- Should the data be stratified or segmented? By location? By territory? By person? By team? By product/product line? This can provide insight into the variation that occurs in the system or process.
- If the data is being stratified, can this be done with different colors, marks or columns on the same check sheet, or does it require separate check sheets?
- What does the check sheet tell you about what action to take or what further information is needed?

When used to count occurrences of things such as reasons for late delivery, types of products ordered or types of customer complaints, a check sheet will provide the basis for developing a Pareto chart. To highlight trends that occur over time, check sheet information will be used to generate a run chart.

Points To Remember About Check Sheets

- Ask the people who will collect the data for their input on creating the check sheet.
- Make the check sheet user-friendly.
- Field-test the check sheet and make modifications as necessary to ensure accurate data collection.
- Decide in advance what to do with the data.
- Provide the results of the data to the people who were involved in its collection.
- Use the results as a basis for taking action.

Work Sampling

Work sampling enables you to get a statistically valid picture of the percentage of time you spend doing various activities without the inconvenience of making a log of everything you do every day. It is especially useful for salespeople and others whose work varies considerably from day to day.

Work sampling, although not definitive, provides sufficient accuracy to be useful. This tool is designed to show how the work processes drive you to spend your time. A simple and non-intrusive way to gather work sampling data is to use a random electronic timer along with a chart such as the following. Electronic random timers, along with copies of the Work Time Analyzer form shown, are available from Conway Management Company.

Work/Time Analyzer©

Position: *Sales Rep* Name: *ALC* Date: *10/23*

Total Obs.	Observations (Check Appropriate Work Activity)	Specific Work Activity	Categories of Work
3	— — —	*Getting Orders*	**Value Added Work**
6	— — — — — —	*Sales Preparation/Anal.*	
2	— —	*Prospect Qualification*	
1	— —	*Sales Training/Meetings*	
	—	*Ident. Cust. Needs/Wants*	
4	— — — —	*Traveling*	**Necessary Work** (Non-Value Added)
1	—	*Record Keeping*	
2	— —	*Reports*	
1	— —	*Expense Accounts*	
3	— — —	*Meetings*	
		Meals	
2	— —	*Complaints: Quality*	**Unnecessary Work** (Rework)
3	— — —	*Complaints: Pricing*	
1	—	*Complaints: Deliveries*	
		Complaints: Wrong Prod.	
1	—	*Complaints: Part Ship*	
2	— —	*Reports (some)*	**Unnecessary Work** (All Other)
1	—	*Filing (some)*	
3	— — —	*Meetings (some)*	
		Analysis (some)	
1	—	*Entertainment (some)*	
1	—	*Traveling (some)*	
1	—	*Personal Time*	**Not Working** Authorized
		Holiday/Vacations	
		Breaks	
1	—	*Waiting Time/Idle Time*	**Not Working** Not Auth.
1	—	*Breaks*	
		Arrive Late/Leave Early	
		Time of Observation	

Total Observations = 41
29% Value Added
27% Necessary Work (Non-Value Added)
17% Rework
20% Unnecessary Work
7% Not Working

The completed Work Time Analyzer form provides a clear picture of how much time is spent on the various categories of work and the activities in each category.

To get started, write down all of your standard activities across the top of the chart. If several people with the same job or from the same area are involved in the work sampling study, come to an agreement on the activities and the categories in which they fall. This will make your results more useful. Make the categories broad enough so you don't have to constantly add new activities. Typical activities for a salesperson would be:

- Phoning customers to promote sales
- Responding to customer inquiries via phone, e-mail, fax or writing
- Phoning customers and internal departments to resolve complaints
- Planning sales calls
- Learning about new products/services
- Attending a sales meeting
- Writing up an order
- Traveling to meet a customer
- Entertaining a customer

Next, determine how many observations you would like to collect each day. A typical number for sales and customer service is twenty observations in an eight-hour period, or an average of two and one half per hour. Set the Conway Random Electronic Timer for the desired rate, and it will generate a signal (either a beep or a vibration, depending on the mode you prefer) at random intervals. When you receive the signal, record a tick mark in the column indicating the activity you are doing at that moment.

Intellectual honesty and amnesty are necessary for accurate data collection using this method. Don't worry if you seem to be getting a large number of observations for activities that don't seem representative of what you do. Over the course of the study, the large number of observations taken will give you a sufficient degree of accuracy. As a rule of thumb, you should have at least 400 observations over a cycle of your work. When we refer to a cycle of work, we mean a long enough period of time to give you a reasonably accurate representation of your work. For example, sales activities during the last week of the month may differ significantly from those at the beginning or the middle of the month so you will want to design the study to capture the entire month.

Before asking others to sample their work, sample your own work first, especially if you are manager or supervisor. This will give you a feeling for the challenges of classification as well as the experience of recording what you do. It will help you understand how much time is spent on activities that do not add value, despite your efforts to work on the right thing. This will help you to understand and interpret others' results.

Involve participants early when you plan a work sampling study. Make it clear that the purpose is not to find poor performance, but to help the entire organization do more meaningful work. Let people know how you plan to share and use the results of the study. Use the following pie chart to indicate you expect to find a lot of waste. Explain that more than 90% of problems are caused by the process or the system. The work sample will demonstrate how the system drives you to spend you time. Only if people perceive it to be in their interest to cooperate and share information without fear of reprisal will they do so.

The Way We Work

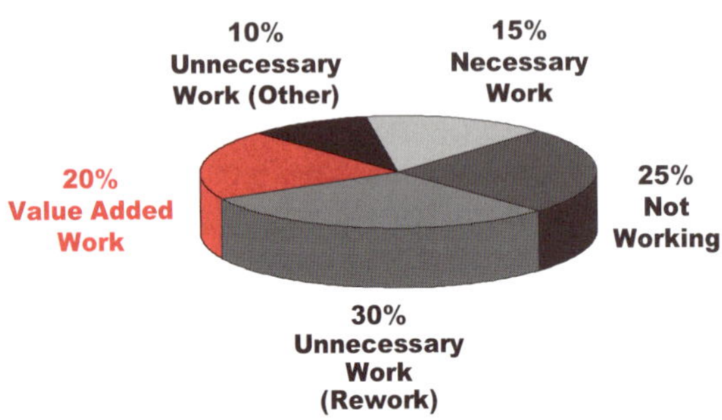

One sales manager collected all of his salespeople's samples and developed a composite for the group. He then gave each salesperson a summary of his or her work sampling study along with the group composite. The group analyzed and used the composite to identify issues to go to work on collectively. Privately, each individual compared his or her result to that of the group to determine what they could do on their own to make improvements in their use of time.

To get the results of the study, add up the number of observations for each activity. Convert this number to a percentage by dividing it by the total number of observations. After the results are accumulated, make a Pareto chart of the work categories. This will give you an initial idea of how people's time is spent. Ask these questions: "Do I like what the Pareto tells me? What percentage of the time is spent on value-added work?" "If everything were right, what would this Pareto chart look like?" Then make a Pareto chart of the specific activities to help determine your priorities for eliminating waste.

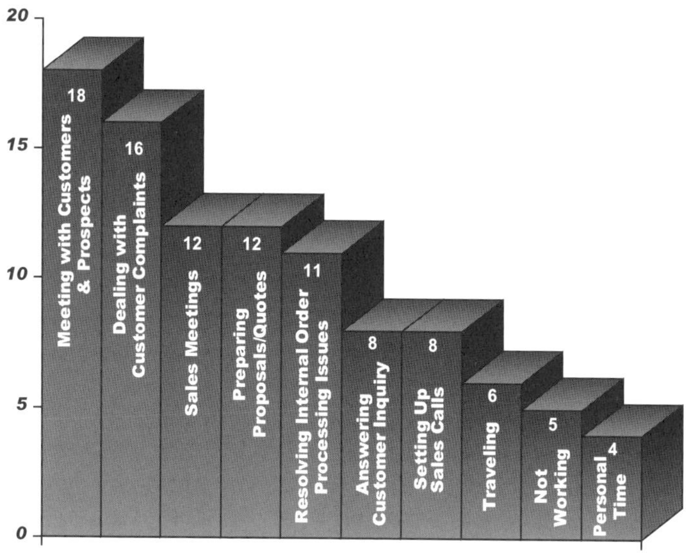

The breakdown of work for salespeople can be a real eye-opener. People typically assume that most of a salesperson's day is spent on value-added work. Once you study the work, you find that much of it does not add value at all! You'll find that too high of a percentage of your time is spent on activities that do not add value to the customer, nor contribute to the overall success of the business.

A good place to start is to look at the time spent on rework and unnecessary work. What obvious activities can be eliminated immediately? Select a specific activity in the category of rework or unnecessary work as the focus for an improvement project. You may need additional work sampling studies to understand the work more fully.

Surveys

Surveys can be extremely useful for gathering information from those directly involved in a process or those directly affected by the outcome of a process. The people who are closest to the work know the most about the waste, problems, and complexities.

As with all data collection, determine what information you are interested in finding out about. Then, prepare a list of questions that concentrate on the way the work is done and the problems caused by the process. Don't get mired in problems caused by individuals or by personality differences. Test the questions by sitting down informally one-on-one with two or three individuals to see if the questions are effective in getting to the waste. Encourage a general discussion of the process to see if there are other things you should be inquiring about, then refine the questions as necessary.

Informal surveys are most helpful if the process involves only a few individuals. For large groups, follow up the selective informal surveys with a formal survey.

The following steps are important if you are to obtain honest, open responses and to maintain the support of those being surveyed:

1. Explain the purpose of the survey to the participants.
2. Tell them what you plan to do as a result of the survey.
3. Communicate the results of the survey to them promptly.
4. Develop an action plan based on what you learned from the survey.
5. Follow-through with your plans and share the results with the participants.

Surveys of external customers should be designed not only to detect problems but also to determine how you can better meet current and emerging customer wants and needs. Be prepared to respond promptly to any concerns customers express. Rather than surveying all of the customers at once, you may want to start with a few accounts to see how well the survey questions and the methodology works.

Questions To Ask Before Conducting A Survey

- What do you want to find out from the survey?
- Who are the people who can give you that information?
- What is the best way to communicate the questions to survey participants?
- Are you prepared to respond to concerns uncovered by the survey?

Points To Remember About Surveys

- The purpose of the survey is collect important information to let you know how you are doing and to identify opportunities for improvement—not to find individuals to blame.
- Design survey questions carefully so that you are able to find out what you what to know.
- Communicate results to participants promptly.
- Take action on the results of the survey.

Charting Techniques

Once you've gathered data, you need a way to analyze it, interpret it and communicate it. The simple charting techniques covered in the next section provide you with ways to organize the data so that you can see important patterns or trends. They enable you to look at data objectively, communicate the information to others and to drive your improvement efforts. They are the language of continuous improvement.

Pareto Charts

A Pareto chart is a simple bar graph that shows the frequency of the causes, sources, types, or reasons for problems and/or opportunities in ranked order, from largest to smallest. This ranking helps you to see the relative magnitude of the factors measured and to determine what to work on.

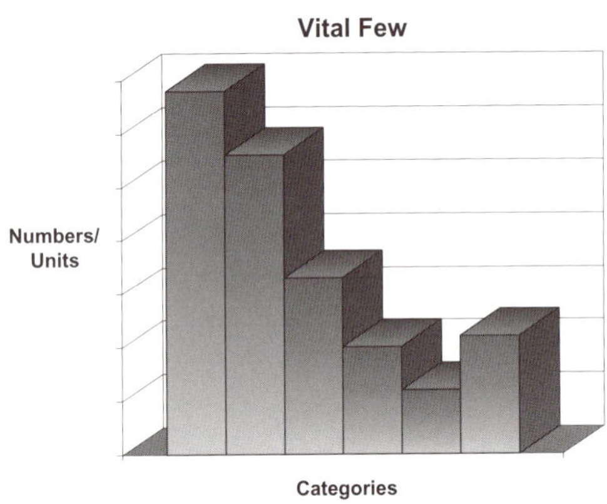

Pareto charts represent a way of thinking called the ***Pareto mentality.*** The most important decision you make each day is what to work on. The Pareto mentality keeps you working on the most important items.

The Pareto Principle is based on the "80-20" rule that says that 80 percent of the troubles, problems and waste are caused by 20 percent of the underlying factors. Working on the "Vital Few" things that cause most of the problems will yield the biggest improvement for your investment of time and resources. This principle applies to proposal errors, customer complaints, anything that involves work and is important to understand. Although it is desirable to work from hard data, you can use estimates or opinions, such as estimated sales for new products, where facts are unavailable.

The Pareto mentality also ensures that you don't just pick the largest column on the chart to work on. You may need several Pareto charts to help you decide on the right thing. Ideally, you want to work on those things that yield the biggest benefit for the least amount of time and dollars. This would lead you to make a Pareto chart of each category by frequency, cost, time to do or any other important consideration.

Making A Pareto Chart

Pick an important area of waste or opportunity and decide how to measure its various elements—number of errors by type, dollars spent by account, sales dollars by territory, time taken by activity, etc.

- Assemble the data you have collected in check sheets or other reports. Make sure you have enough data to accurately represent the relative magnitude of the various elements. Count the frequency or quantity of each type, cause, reason or source, dollar amount or percentage.

- Construct a bar graph with the frequency or quantity shown on the vertical axis. Be sure to create a scale that will accommodate the largest column and that accurately reflects the comparative heights of all the columns. Draw the horizontal axis such that it is long enough to accommodate all of the columns.

- Start with the category with the greatest frequency or quantity. Draw a column on the left side of the chart at the height representing the frequency or quantity. Add the other categories in descending order, drawing a bar (column) for each cause at with the appropriate height. Items that occur infrequently can be grouped in an "other" or "miscellaneous" category. Keep that bar on the far right, even if it's not the smallest bar.

Label your chart with the following information:

- What the chart is
- Units of measure (vertical axis)
- Name of each column (horizontal axis)
- Period of time the data in the chart represents (day, week, month, quarter, year, etc.)

The following Pareto chart was prepared by a client hotel from the customer complaints received in one month:

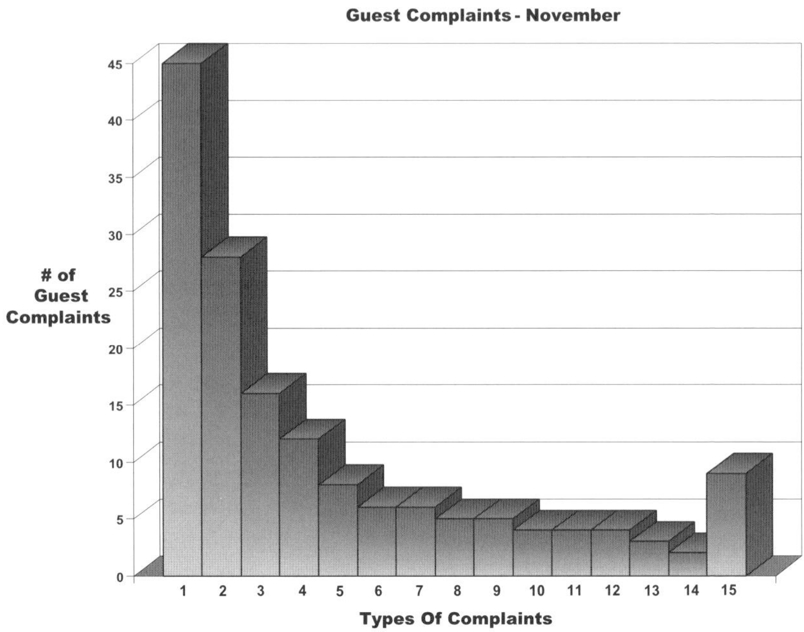

1 Noise
2 Parking
3 A/C Noisy
4 Small Room
5 Inattentive or rude employees
6 Lack of amenities in room
7 Maids disturbing
8 Charged for local calls
9 Mini bar
10 Housekeeping issues
11 Reservation wrong
12 Not warned about construction
13 Bed uncomfortable
14 Shower does not work
15 Other

This chart shows that noise—both internal and external—was by far the most common complaint. Prior to tracking and summarizing customer complaints, the hotel assumed the main problem was a housekeeping issue and had focused its improvement efforts in that area. Once people saw the Pareto chart, they knew they needed to change their focus.

Using Pareto Charts

Your Pareto chart will help point to areas for improvement, but it's important to chart the right things. Think carefully about the factors that influence your decision about what to work on and display them in Pareto charts when possible.

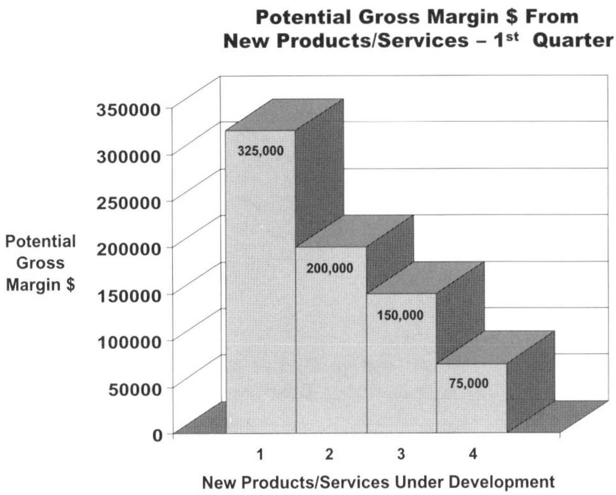

This Pareto chart could help in the allocation of limited resources. Notice that the opportunities are measured in gross margin dollars, not in sales dollars. A chart in sales dollars might be useful as well, but the product with the most sales potential might not be the most profitable.

The next chart shows the number of successful closing appointments in May for products A, B, C, & D.

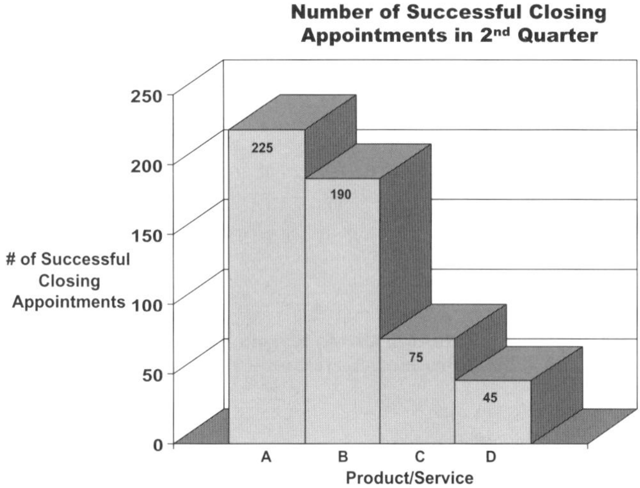

Next, if the chart is changed to show average gross margin dollars, you see a very different picture.

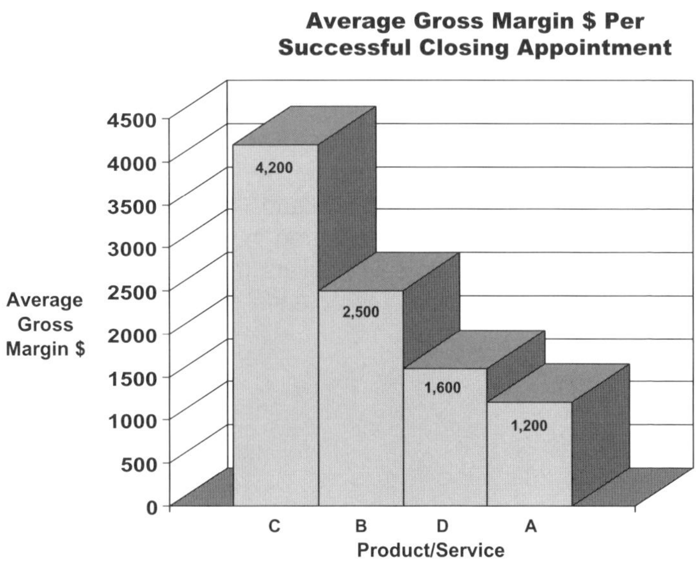

Product A, for which there are the highest number of successful closing appointments, yields the fewest sales gross margin dollars. Product C, which has relatively few successful closing appointments, yields the most gross margin dollars per close. You can conclude from this that it would be worth your while to increase Product C closings.

Often when you start to work on the issues represented in your Pareto chart, you find that you need more detailed information. You can make a sub-Pareto chart of the elements that make up each column. For, example, you may discover there are several different reasons for unsuccessful closes. You can then make a Pareto chart for those reasons.

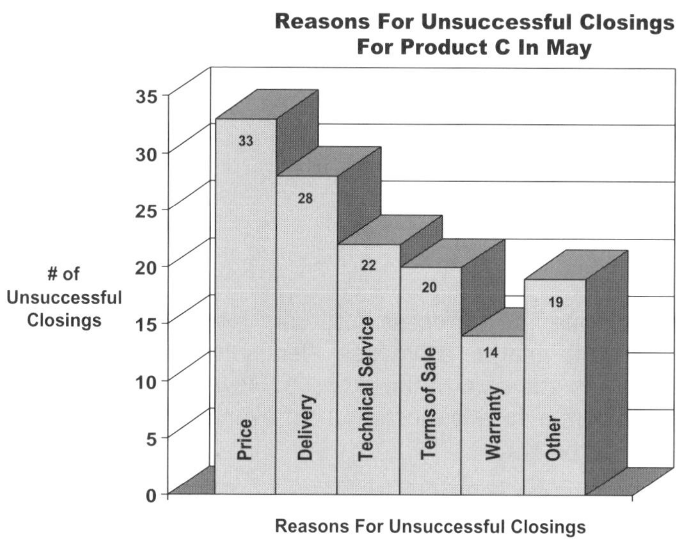

In this case, you may need to investigate the reasons for pricing problems, or the reasons for delivery problems.

Go down as many levels as necessary to arrive at the fundamental causes. When you dig deeply enough, it often becomes obvious what must be done to fix your problem.

Displaying important data in Pareto charts makes it easy to see what is going on and helps you to continually evaluate what is most important to accomplish. It helps you get the facts to prevent you from spending time on minor problems or opportunities while neglecting important ones. Even when you don't draw a Pareto chart, visualize one for every area of your activity. The Pareto mentality is a distinguishing characteristic of people who effectively work on the most important things, not just the urgent ones.

Points To Remember About Pareto Charts

- Pareto charts are a tool to help you prioritize information in order to decide what to work on.
- Use the Pareto mentality to keep focused on the "Vital Few" opportunities and not the "less important many."
- Make as many Pareto charts as are useful for effective decision-making.
- Arrange the information you have gathered in bars in descending order, except for the category "other." It is always at the far right even if it is not the smallest bar.
- Study the major categories and make sub-Pareto charts of each, if necessary, to understand the process better.
- Keep breaking down the opportunities, issues or causes further and further until what to do becomes obvious.

Flow Charts

A flow chart is a map of a process. You can use a flow chart to document the work, analyze a process, identify waste, plan improvements, and standardize the process. Flow charts also serve as training and communication tools to help everyone visualize the process in the same way.

A flow chart can be constructed with varying degrees of detail depending on how you plan to use it. The simplest type, a block diagram, shows the major steps in the process in sequence.

Making A Sale

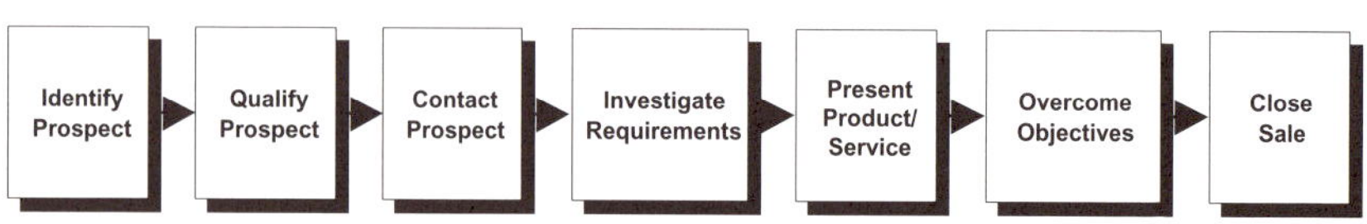

This broad overview of the selling process could be used to outline the basic steps in the process and to begin a discussion of the work. But usually more detail is needed if your purpose is process improvement. The most common flow charts show more detailed steps with an estimated time for each step as well as the total elapsed time. If the process is unusually complex, each step may be broken down further into smaller steps as needed.

Making A Flow Chart

More detailed flow charts break the process down into finer steps that more completely describe the work. This is where you begin to see the troubles, problems and waste. In addition to listing the steps, detailed flow charts use symbols for each step to let you know the type of work activity the step represents. The following six conventional symbols are usually sufficient.

The Symbol	Denotes	Examples
Diamond	Decision	• Call prospect today? • Does product B12 meet prospect needs?
Square	Control/ Inspection	• Verify commission to be paid • Check sales order for completion
Circle	Operation	• Call prospect • Write letter
Arrow	Movement/ Transfer	• Take goods to customer • Travel to customer location
D	Delay	• Wait for approval on customer terms
Triangle	File	• File customer correspondence

Steps in making a flow chart:

1. Name the process and describe its inputs and outputs, its suppliers and customers.

2. Decide on the boundaries. Where will the process being studied start and end? What operations will be included? What operations will be excluded?

3. Decide on the level of detail that will be useful. How much detail will be necessary to understand the work and find the waste?

4. Observe the process, talk to the people doing the work and, if possible, involve them in developing the flow chart. If it is a new process being planned, a group discussion to visualize it is helpful.

5. List the steps in the order in which they occur. Be sure to include rework, variations to the usual process and delays. Post-it Notes™ can be used to list one step per note. This makes it easy to make changes or to add more detail.

6. Add the appropriate charting symbol to each step.

7. Connect the symbols with lines showing the direction of the work flow.

8. Add the time for each step and the time between steps to your chart. Add up all of these times to determine the total cycle time for the process.

9. Share the flow chart with others, reviewing it for accuracy. Make revisions if needed.

The following example shows a sales process.

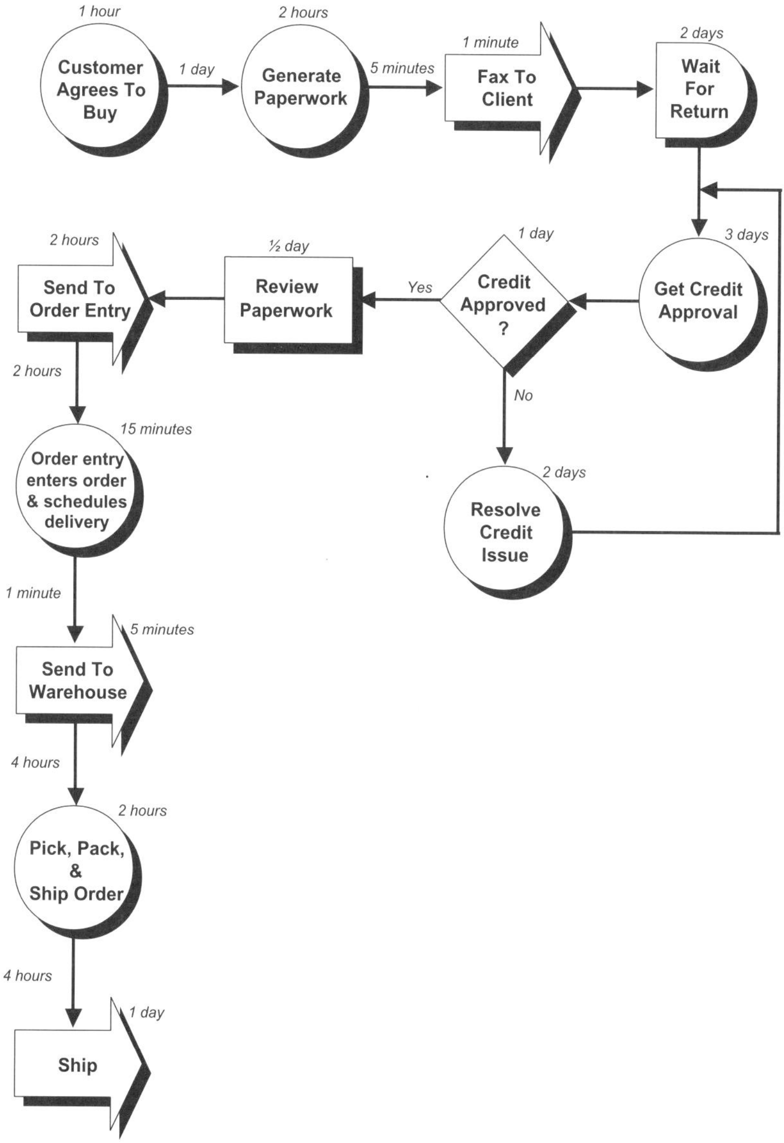

You can begin your analysis of the flow chart by making a Pareto chart to show the time spent on the various types of activities in the process.

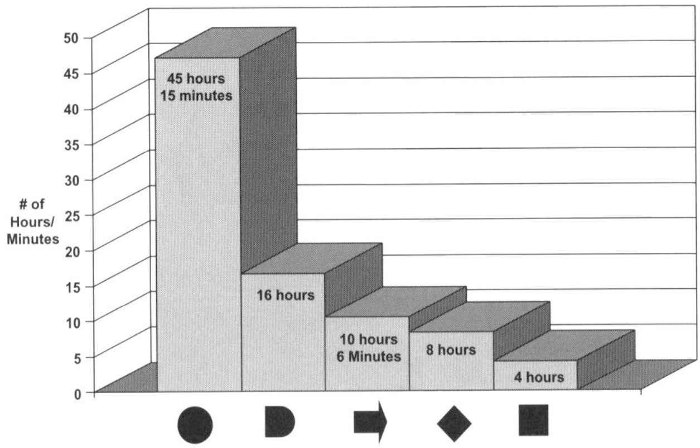

Once you have prepared your flow chart, use this process to search for waste:

- Collect data about numbers of occurrences and reasons for rework or other problems.
- Look at all the steps, such as delays, that do not add value and find reasons for them, then try to find ways to reduce or eliminate them.
- Identify specific areas for improvement (rework loops, delays, multiple inspections or authorizations, etc.)
- Imagineer the perfect process and compare it to the current process.
- Determine if you can change, rearrange, combine, or simplify to improve the process.
- Make sub-process flow charts to break the work into finer pieces as necessary to understand the process fully.
- Draw a new chart with the modifications you want; discuss it with the people involved; and test it.

Questions To Ask About The Process

- How well does the process meet customers' needs and requirements?
- How long does it take? How long should it take if everything were right?
- How much does it cost? How much should it cost if everything were right?
- What works well about the current process?
- What does not work well?
- How does your process compare to your competitors' or the "best in class" for effectiveness and efficiency?

Questions To Ask About Each Task or Activity

- What is its worth to the customer?
- How does that compare to its cost?
- Is the purpose of the task still valid?
- What risks would eliminating the task pose?
- Is it being done by the right person or department?
- Can you transfer it to someone outside the company?
- Can you consolidate it with another job within the company?
- Can you reduce the frequency of the service?
- Can you reduce the content of the service?
- Can you reduce the number of people or groups receiving the service?
- How can the task be improved?
- Can you change the methods to make the job more efficient?
- Can you mechanize it or computerize it?
- How much time is spent waiting for work?
- Can you level the workload and/or vary the work force as needed?
- What can you do to understand the process better, eliminate causes of non-value-added work, and improve the quality, cost, and effectiveness of value-added work?

The following Work Improvement Checklist will help with your process analysis. Look for improvement opportunities in the following order:

What Can You Eliminate?

- Does each step add value?
- Is there duplication?
- Can you eliminate delays? Inspections? Filing?
- Can you eliminate transport steps?
- Are some steps rework?

What Can You Change?

- How can the process be changed? Different methods? Different technology?
- Is there a less costly service?
- Can you reduce the frequency of the service?
- Can you decrease the number of people it takes to provide the service?
- Can you reduce the time it takes?

What Can You Rearrange?

- Is the sequence the most efficient and effective?
- Can steps be done in parallel instead of sequentially?
- Can the steps be done by someone else to improve the quality, cost or delivery?

What Can You Combine?

- Can any steps be combined? With a supplier's operation? With a customer's? With another department's? With another person's?

What Can You Simplify?

- What is the simplest way to achieve the objective(s) of the process?
- Are instructions easy to understand?

Imagine

- What would the perfect process look like?
- How much time would it take?
- How much would it cost?
- What would customers' reactions be to the perfect process?
- How would people doing the work feel about the process if it were perfect?
- What is the gap and the cost associated with it?
- What would it be worth to close the gap?

Points To Remember

- Start with a simple chart and add detail as needed to help you find the waste.
- Classify the activities using the six conventional symbols.
- The people doing the work should chart the process.
- Collect data to verify the flow chart and to learn more about the reasons for troubles, problems and waste.
- Ask the work improvement questions.
- Draw an imagineered chart based on the way you think the process should work.
- Compare the differences between the current and imagineered process.
- Decide how to make the changes that will get you closer to the imagineered process.

Run Chart

A run chart plots data points of key process variables in order as they occur over a period of time. It shows the amount of variation your process produces and highlights trends and shifts in the average level of performance. Run charts provide visual clues about how the process is operating, and whether it is improving, deteriorating, or staying the same. Run charts are the most frequently used tool for tracking performance.

A run chart has two axes. The vertical axis shows the units of measurement. The horizontal indicates time or sequence of the data points.

$ Sales By Salesperson B

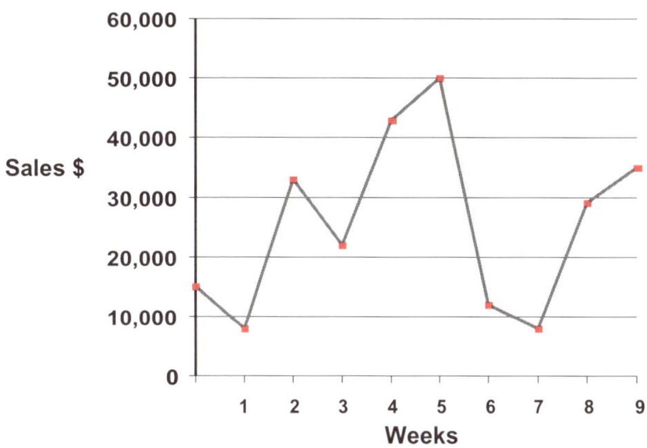

Making & Using A Run Chart

Selecting the right elements to measure that accurately reflect what is happening in the process is one of the most important decisions you make. The measured element, often called a key performance measure, should be critical to the process. Choose process variables that affect the effectiveness and efficiency of the process.

Sometimes more than one unit of measure is necessary to understand what is going on. For example, knowing the number of returned phone calls is important but it is also useful to know the percent of phone calls returned. You might take very different actions in response to either measure but together you get a more complete picture. Keep in mind that measures are more powerful than counts in providing useful information for analysis.

"What gets measured gets managed." Measures are a reflection of what is important to you so people pay attention to what gets measured, sometimes at the expense of other important factors. For example, if you are measured strictly on gross revenue produced, you may ignore gross margin dollars and contribute to a loss in overall profitability. Make sure they are the right things to focus on to produce the best overall result.

When you decide on the charts you believe will be useful, try to get some historical data to give you some baseline information. This will show you the current operating level and variation in the process. If no historical data is available, set your chart up and plot the data points as they present themselves.

Update charts as soon as the data becomes available. Timely information is far more useful than out-of-date statistics and it is critical if special causes of variation are present. Regularly updating and analyzing your key performance measures keeps you focused on the most important things and increases your effectiveness in making data-based decisions.

To construct a run chart:

- Draw a vertical axis and mark the scale with whatever unit you are measuring (percentage of unsuccessful closings, number of demos, sales dollars, etc.). Set the vertical scale to cover the range you expect and in such a way that, when data points are plotted, they will neither overstate the variation nor understate it.

- Draw the horizontal axis and mark it to indicate the frequency at which data will be collected (daily, weekly, monthly, etc).

- Plot the data in order as it becomes available. Be sure to determine how much data over what period of time you need in order to get a good picture of how the process is operating.

- Once you have enough data points, calculate the average and draw a horizontal line on the chart to indicate the average.

Using A Run Chart

A run chart is an excellent way to express data in context. In analyzing the data, look for overall patterns rather than reacting to individual data points, unless a truly special cause of variation has occurred. Special causes may be present if you see the following patterns:

- A shift in the process average, when you have 9 or more points in a row on one side of the center line or average.

- A trend, when you have 6 or more points in a row increasing or decreasing

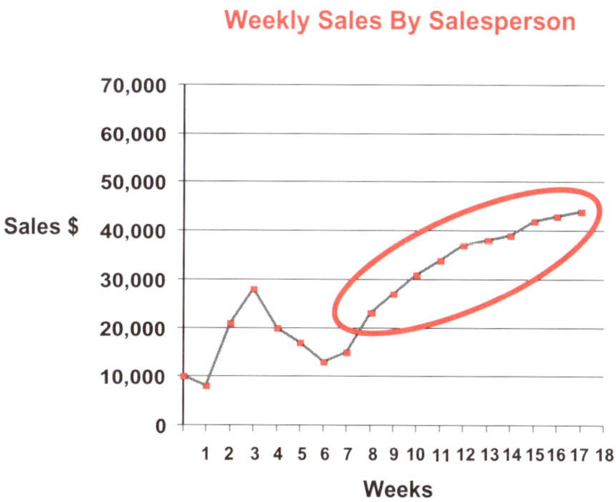

- An unusual event, when you have a data point that falls more than 3 standard deviations from the center line.

Weekly Sales By Salesperson

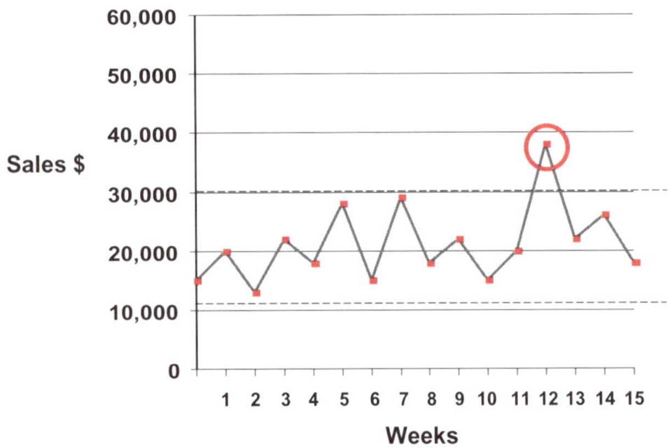

When you see that these special causes are occurring, make sure you track down the reasons for them. If the trend or shift is in the right direction, try to make the changes that caused the improvement a permanent part of the process. If the trend or shift is in the going in the wrong direction, track down the reasons and try to reduce or eliminate them.

- Sometimes it is helpful to compare two similar processes by showing them on the same chart:

Weekly Sales By Salesperson

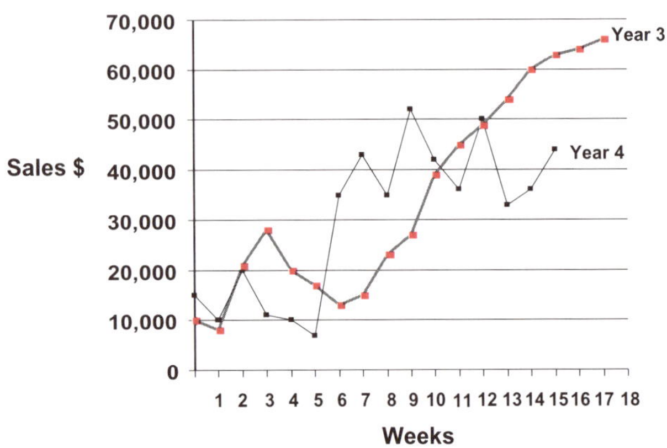

Questions To Ask About Run Charts

- Where is the average level for the process?
- Do you like the current level?
- What is the cost to operate at that level?
- If everything were right, at what level would the process be operating?
- What are the reasons you are at the current level instead of the desired level?
- How much variation is there around the average? In other words, how predictable is the process?
- How much variation would there be if everything were right?
- What is the cost of having that much variation? What is the impact on customers?
- What are the reasons for the variation?
- Are any trends showing up? Has there been a process shift?
- If so, is it in the right direction? If it is, what caused it and how can you keep the improvement? If not, what caused it and how can you reduce or eliminate the reason?
- Does this process have a cycle to it?
- What other charts do you need to understand and improve the work?

Points To Remember

- A run chart shows what happens to a key performance measure over time. Make sure the data is plotted in time-ordered sequence.
- Time or sequence goes on the horizontal axis; count or measurement on the vertical axis.
- Always show what is being measured, when it was measured, and any other information that will help the chart reader understand what is happening. Make notes on your chart when something special or unusual has occurred or when you make a change. These notes will help you to better understand what the chart is telling you.
- Update the chart as soon as the data is available.
- Choose the right things to measure and an appropriate scale to give you the best picture of what is happening.
- What you choose to chart shows what you believe is important. Chart only those things that matter most to the effectiveness and efficiency of important processes.

Fishbone Chart

A fishbone chart, also known as a cause-and-effect diagram, shows relationships between causes and a particular effect you want to study.

Unlike the other charts used in The Right Way To Manage©, a fishbone chart begins with opinions, rather than facts. It represents an opportunity to find out from the "experts"—the people who know the most about the work—the possible reasons for a specific problem. Once you have the opinions, you'll want to verify them with facts before taking action to address the problems.

- The first step in preparing a fishbone diagram is to brainstorm the major cause categories that will help people think creatively about all the specific reasons for the problem in which you are interested.

The common classifications of causes are people, materials, machines, and methods. Information, measurement, management and environment are also relevant in most situations. When analyzing administrative processes, "policies" and "procedures" may be useful major classifications. Any major category is suitable if it helps people think creatively about the cause of the effect. Use only those topics that make sense and will point in the direction of many potential causes.

The major categories of causes are assigned to the large branches of the diagram.

A simple cause-and-effect chart looks something like the skeleton of a fish, hence the name.

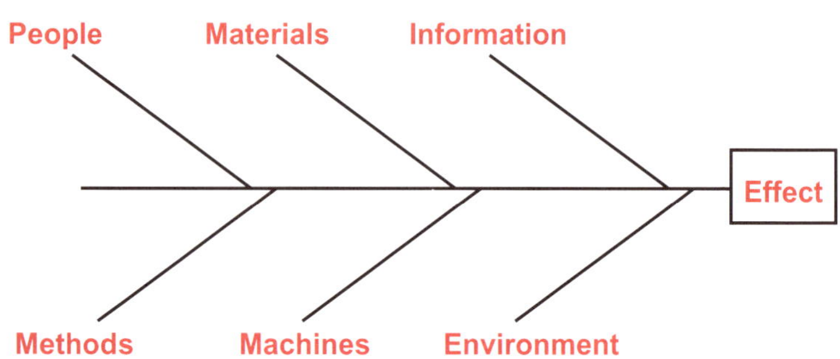

Specific causes that fit in the major categories are brainstormed and added to the diagram. The chart usually becomes more complex as more specific causes are identified. It begins to resemble a tree lying on its side with main branches, sub-branches, twigs, sub-twigs, etc. Sometimes a root cause may be found on a sub-twig.

Making & Using A Fishbone Chart

- Decide what problem or opportunity to study. Clearly define the effect and place it in a box on the right side of your diagram.
- Gather people who have knowledge of, or experience with, the effect you are studying. Try to assemble a diverse group in order to get varied or multiple perspectives. Although an individual can create a fishbone chart alone, a group generates more ideas for possible causes.
- Select a leader to manage the brainstorming session and a scribe to record the ideas on the fishbone diagram. Use good brainstorming techniques to decide on the major cause categories and then to generate as many possible reasons for the effect as possible.
- Once the ideas have been listed, clarified and discussed, determine the most probable causes of the effect and decide what data you need to collect to verify them.
- Create the environment where people feel free to tell you the truth about what is going on without fear of repercussions. Remind people you are focusing on the process, not looking to blame individuals, so that you can find and eliminate waste.

Brainstorming

The process:

1. Clearly state the idea to be brainstormed.
2. Ask people to brainstorm individually and write down their ideas silently.
3. In round robin fashion, share everyone's ideas with the group.
4. Review each idea:
   —For understanding and clarification
   —To combine similar ideas and eliminate duplicates

Ground Rules For Group Brainstorming Are:

- Record all ideas accurately
- Share one idea per turn in a round-robin style
- Be open to new ideas; don't criticize or comment
- Remember that no idea is "too wild"
- "Piggyback" or build on others' ideas
- It is okay to pass when you have no more ideas to contribute

Try to have everyone contribute ideas about possible causes. Restrict the brainstorming to **causes**, **not solutions**. The scribe should record the possible causes as they are suggested. Ideas may be recorded directly onto the fishbone diagram, but it may be difficult to classify them properly at this point without slowing the flow of ideas. An effective way to avoid this problem is to write each idea on a Post-it note and construct the fishbone later by posting the notes in the appropriate place. A sample fishbone from one such group is shown below.

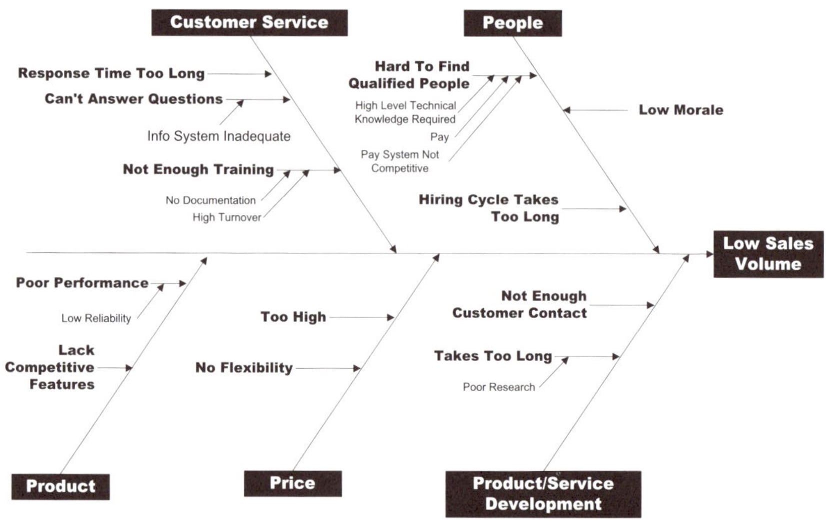

After constructing your fishbone chart, have the group decide which causes they believe have the greatest impact on the problem or opportunity, by using either consensus or a vote. Now that you have priorities, you will probably need hard facts. Ask team members or others outside the group to gather initial data. As always, fix the obvious problems right away.

Once you have collected the data, make a Pareto chart of the major causes and review it as a group.

For each of the major causes, you now want to delve into the work by asking "Why?" several times.

For each cause you identify, ask what are all the possible causes of that reason. This will get you moving in the direction of identifying the root cause of the problem.

Here is an example of using the 5 Why's with the fishbone chart. The effect is decreased sales and the cause category is Customer Service.

Customer service response time is too long.
Why?

The call center receives heavy volume of calls that require research or manual checking.
Why?

There is a lack of documentation to help customer service reps make decisions quickly.
Why?

There was a major change in procedures without documentation.
Why?

We thought that making these changes would save time. Now we have heavy turnover in the department and do not have the resources to complete the documentation.
Why?

It is a frustrating department to work in without documentation and our people are transferring to other departments.

Until the organization stops "just working on the waste" and spends the time to provide the documentation and training to the reps, the problems will continue.

Put the fishbone chart in a final form, highlight the major causes, and make copies for all participants. Consider placing the chart on a bulletin board as a way to solicit further ideas.

Begin to work to study and reduce the major causes and schedule further meetings as appropriate.

Questions To Ask About Fishbone Charts

- Is the problem or opportunity being studied important?
- Is it clearly defined?
- Do you have the right people involved in generating ideas, people who understand the work and the problem?
- Does the atmosphere make people feel free to contribute their ideas?
- Have you identified all of the possible reasons for the problem?
- Have you asked "Why?" several times for the most probable causes?
- What data is needed to verify each of the major causes?
- What action should you take or what teams should you form to reduce or eliminate the problem?

Points To Remember

- Categories represent the major elements/contributors to the problem, process, or opportunity.
- Often-used categories for a fishbone are People, Materials, Methods, Machines, Environment, and Measurement.
- Involve people from various aspects of the process who have knowledge of the work, including all the troubles and problems. Specialists outside the process may also help.
- Brainstorm only causes, not solutions.
- A fishbone represents opinions. Verify opinions with facts before taking action.

Histograms

A histogram is a frequency distribution that gives you a shapshot of the variation in a process or a product. It shows how wide the range of the variation is, where it is centered, and the shape of the distribution. When compared to customer requirements or specifications, it illustrates how capable the process is of meeting them. Displayed as a bar graph, it gives you an at-a-glance picture of the frequency of each measurement.

The following histogram shows the response time to customer inquiries for customer service representatives in a transportation company.

Response Time To Customer Inquiries

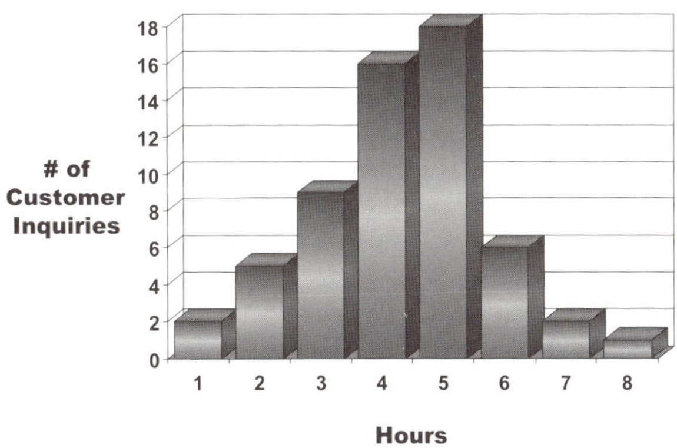

This chart shows that the average time to respond to customer inquiries is just over 4 hours, and it ranges from 1 hour to 8 hours. If the company's objective is that the average response time not exceed 5 hours, it would appear that the company is meeting its objective. However, the average length of time alone does not provide enough information. The histogram reveals that 9 customer inquiries or 15% of the total number, exceeded the company's objective for response time. This histogram gives an idea of the size of the problem and some hints of what needs to be done to fix it.

The next important step is to understand why there is the variation. Sometimes the inquiry is handled in 1 hour, while other times it is 6, 7, or 8 hours. What is the difference? Is response time the right thing to measure? Is 5 hours the right target? A further step would be to analyze the types of inquiries the customer service reps are fielding. Are they the best people to handle these calls? What is happening in the systems and processes that create a need for customers to call?

Constructing A Histogram

The simplest way to construct a histogram is from a check sheet of data.

1. Count the number of data points. You need at least 50 random measurements for a useful histogram.

2. Determine the range by subtracting the lowest value from the highest.

3. Determine the number of class intervals or bars. It depends on the number of data points and is approximately equal to the square root of that number. Here is a general guide.

Data Points	Number of classes
Under 50	5-7
50-100	7-10
100-250	10-12
Over 250	12-20

4. Determine the width of the class intervals. Divide the range by the number of class intervals and round off to a convenient number. Make sure that the intervals are mutually exclusive so that there is no confusion about where a data point should be recorded.

5. Construct the histogram from the available data.

You can approximate a curve with the amount of data you have. If you took more and more readings and added more and more classes, the difference in the bar heights would become smaller until the histogram eventually became a curve.

Using A Histogram

The histogram gives a good picture of the central tendency (average) of the data and the dispersion (variation). It also shows the range of the measurements and the level of performance that the process is capable of delivering (process capability). You can add customer specifications to the histogram. This will allow you to compare the specifications to your process performance to determine whether it is capable of satisfying them 100% of the time. The histogram also shows where the process is centered. Is the process correctly aimed to deliver the results that customers expect?

The shape of the histogram provides additional clues about what is happening in the process. Most processes tend to exhibit a "normal" distribution or a bell-shaped curve.

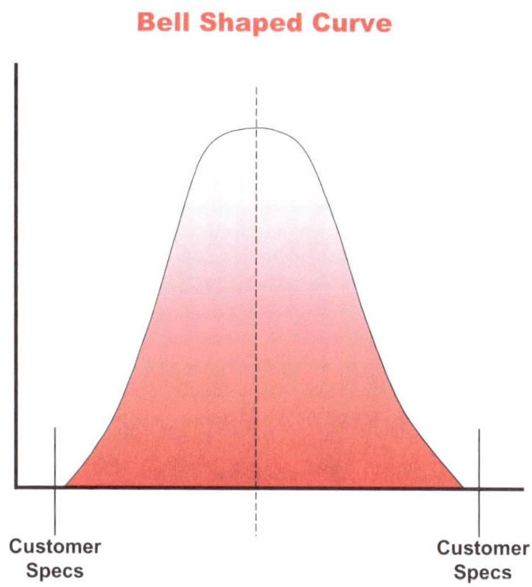

Example of a normal distribution where all of the data points fall within customer specifications

When the histogram appears to end abruptly on one side, it is sometimes called cliff-like. This may be caused when the minimum value at the "cliff" is zero. In manufacturing processes, when the minimum is not zero and there are few if any data points outside specification limits, it is likely that some of the items below the "cliff" have been removed by inspection because they did not meet specifications. This kind of distribution from a supplier indicates the supplier's process is not capable of meeting specifications and is generating waste. If the supplier's item is important, it may be worthwhile to help the supplier to improve the process and reduce the waste, or to examine your specifications to see if they can be altered. You may have a "cliff" on either or both sides of a distribution.

A histogram might also show two peaks instead of just one. This type of histogram reflects two different processes, such as sales from two different territories or time to resolve customer issues in two different locations. In order to effectively use the information, you should segregate the data by process and make two histograms. Then investigate the causes for the differences and improve one or both of the processes.

Proposal Response Time In Days

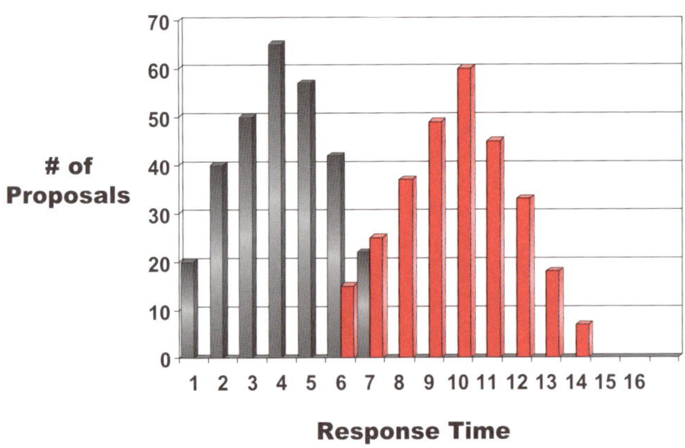

The key questions to ask about a histogram are usually:

- What are all of the reasons for the variation?
- What are all of the reasons the average is where it is versus the ideal?

Brainstorm all the reasons for excessive variation in the process or for the process not being aimed correctly and show them on a fishbone chart. Then collect data and identify what must be improved to reduce the spread of the variation or change the aim of the process permanently.

Questions To Ask

- What is the range of the data? How does the range compare with the specifications? Is the process capable?
- Is it centered where you want it to be?
- How much improvement is necessary? How much is possible? What can you do to improve the aim and narrow the range?
- What is the shape of the histogram? Is it symmetrical? Bimodal? Cliff-like? Skewed to one side or the other?
- What does the shape tell about the process? Does it show a normal distribution? If not, ask why.
- Does the target value for the process coincide with the highest bar and the average? If not, what must you do to shift the process toward the target?
- What other tools would help you to understand and improve the process? Fishbone? Run chart? Pareto chart?

Points To Remember

- A histogram is useful for data that is either counted or measured.
- Gather at least 50 data points. More points may make it more accurate and meaningful.
- A histogram shows the range of the variation and where the process is centered.
- Customer specifications can be added to determine how capable the process is of consistently meeting customer requirements.
- The narrower the range of variation while centered at the target, the more predictable your process and the less waste it produces.
- After process changes, collect new data and make another histogram to show progress.

Correlation Charts

A correlation chart shows the relationship between process variables. The relationship can be either positive or negative. In a positive correlation, as one variable increases, the other increases as well. In a negative correlation, as one variable increases, the other variable decreases.

A correlation chart is used to discover and document possible cause-and-effect relationships based on data (as opposed to the opinions used in fishbone charts). The correlation chart lets you test for a possible relationship between two variables and shows the strength of that relationship. Often you have a number of potential causes for process variation and want to determine which of the causes have the most effect on—or correlation with —the outcome. Of course, showing a relationship between two variables does not necessarily prove cause and effect. A third cause may affect both variables. Use common sense and additional tests, if necessary, to determine whether an apparent cause-and-effect relationship is logical.

In the example below, it appears logical that the market share in the region is somewhat dependent upon the number of sales people in that region. There may be other factors, such as the number of competitors operating in the region, the competence of the salespeople, the amount of advertising, etc. In this case, the number of sales people and the market share show a positive correlation. In other words, as the number of sales people goes up, the market share goes up. If your data points form almost a straight line, it indicates a strong, or positive, correlation.

Correlation charts are most useful when you are uncertain about the causes of variation in a process. Testing the correlation of various factors can help pinpoint the crucial factors.

Making A Correlation Chart

Pick two factors to test their correlation. Since correlation most often implies cause and effect, pick a factor you are trying to control or change (the effect). Then pick a second factor (the cause) that you believe makes the first factor vary.

Gather data in pairs. Let's assume that you are concerned about the monthly gross margin sales people with less than two years of experience are getting and want to determine the need to shift people to territories with fewer new sales people.

Try to gather at least 50 pairs of data.

Prepare a chart with X and Y axes. Plot the expected cause (percent of less-experienced sales people in the region) on the X or horizontal axis, and the effect (sales or market share) on the Y or vertical axis. For ease of reading, the two axes should be about the same length.

Plot each data pair as a point on the chart. If more than one pair falls at the same point, draw one or more circles around the point as needed.

This chart indicates a negative correlation in that with fewer new people in a region, gross margin $ are higher. As the percent of new sales people increase, (perhaps making it more difficult to train and supervise) gross margin $ decrease.

The correlation chart is most useful when a number of variables may affect the outcome of a process. By charting each variable against the outcome or effect, you can see if and to what degree the two factors are related.

The degree of correlation can be determined roughly by looking at the chart. The closer the pattern of dots is to a straight line, the stronger the correlation. A completely random pattern indicates no correlation. A weak correlation indicates that other factors are affecting one or both variables.

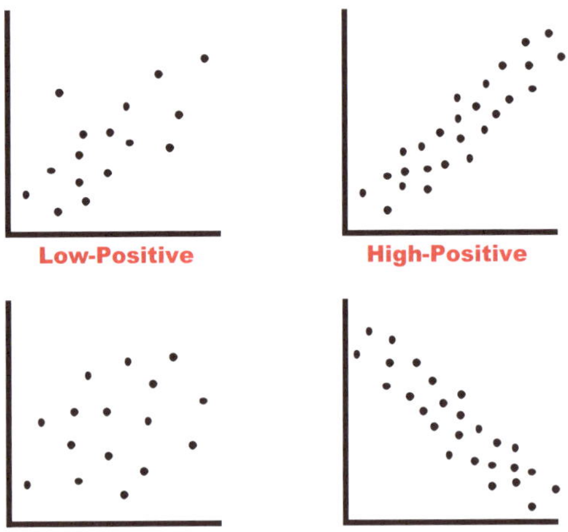

Questions To Ask

- Have you discovered the most important variables affecting the outcome of the process?
- Is there a cause-and-effect relationship between the variables charted, or is it possible that a third (or fourth) factor is responsible for the variation in both?
- How strong is the relationship?
- What can you do to improve the cause in order to get a better result?
- If the correlation is weak, what other factors are involved?

Points To Remember

- A correlation chart can show if there is a relationship between two variables and the strength of that relationship.
- The closer the chart is to a straight line, the stronger the relationship.
- Use correlation charts when you want to determine what effect changing one variable in a process will have on another variable.
- A relationship does not necessarily prove cause and effect because the variation in one or both factors may be caused by a third or unknown factor.
- Apply common sense when looking at possible relationships.

Control Charts

The control chart is a run chart to which lines indicating the upper and lower natural process limits have been added to sort random from non-random events. It is then used to analyze, stabilize, control, and improve the process. The control chart is a powerful tool used to define the voice of the process and then characterize the behavior of the process as either predictable or unpredictable. The value of this information is that it provides you with direction on what action to take: either track down the causes of individual data points or study, change and improve the process.

There are many different kinds of control charts so training is required to select the appropriate chart and construct it properly. A good reference is Dr. Donald Wheeler's textbook, *Understanding Statistical Process Control* published by SPC Press, Inc.

The control chart takes two key parameters of a histogram — the average and the range —and plots them over time. Statistical formulae are used to calculate control limits from the data that are then plotted on the chart. These control limits indicate the range of variation inherent in the process. Data outside those limits indicate special causes of variation, as discussed in Section 1. Special causes need to be investigated. Those that are harmful need to be eliminated to achieve a stable process with a known range of variation. Those that are helpful need to be made part of the process.

As in all studies of work processes, the most important step is deciding which factor to study. The control chart is useful for any measures that are important to the organization and for which data can be made available.

Why Is Understanding A Control Chart Important To Sales People?

One Conway client had a quality problem that was affecting sales. It tracked the problem to "pin marks" on the component they supplied to their major customer. The sales person assured the customer that this defect would be quickly eradicated. However, a control chart of the pin marks showed that there were no special causes that could be readily remedied. Instead, the problem was systemic and would require retooling the entire production line.

The point of this story is that the sales person did not understand that the variation was inherent in the current process. Had he understood the nature of the variation, he would not have promised that the problem would not happen again. But he didn't know. He and the customer would find out with the next delivery. The sales person eventually lost the business, but he may have been able to salvage the account if he understood the situation and if there was something he could do to rectify it.

Questions To Ask

What are the key variables?

- What type of control chart is appropriate for the data?
- Is the process stable?
- If the process is stable, are the level and the range acceptable?
- If not, what are the causes inherent in the process that need to be identified and addressed?
- If the process is not stable or predictable, what needs to be corrected to stabilize it?
- Is the control chart being kept on a real-time basis?
- Have people been trained in the use of the control chart so they can react appropriately to the data?
- The purpose of analysis is insight to be used as a basis for action.

Points To Remember

- Select the proper type of control chart for the data you are collecting.
- Plot points on a control chart as soon as practical to discover problems quickly.
- Make sure people using control charts are trained to interpret them and react properly.
- For any out-of-control point on the chart, record the cause and the action taken.
- Compute a new average and range after making process changes. This will tell you what effect the change had on the process.
- Having a stable and predictable process does not guarantee that you are meeting customer requirements.
- Once you have a stable process, study, change, and improve the work to improve efficiency and effectiveness

Prioritization Matrix

The prioritization matrix is a tool to make important decisions in a visible and disciplined fashion. Its value lies in enabling you evaluate your options in light of clearly identified criteria that have been weighted in terms of their relative importance. Effective prioritization is critical to working on the right thing and the matrix shows the logic behind the decision.

The matrix can be used to select opportunities for improvement, prospects to pursue, markets to enter, products and services to offer, team members, solutions, equipment, etc. It is especially helpful when there is a general lack of agreement on a particular issue or when there are many factors to be thoughtfully considered.

Use this tool to:

- Help decide what to work on
- Provide a logical and visible process for decision-making

Making a simple prioritization matrix

- List the items to be evaluated
- Select criteria against which the items will be evaluated
- Determine the rating scale and relative weights of criteria
- Rate each item against each criterion, and multiply by the weighting
- Total the ratings
- Pareto-chart the results
- Do a “gut feeling” check on results
- Decide what to work on

An example of a simple prioritization matrix is shown below. The Dogwood Company wants to determine which prospect they should pursue first to increase their revenue.

They decide that:

- Likelihood of Success is the most important factor and assigned a weight of 2.5
- High Dollar-Volume Potential with longer term business potential is next in importance and assigned a weight of 1.50
- Cycle Time to Close is slightly less important and assigned a weight of 1.0

The weight will be multiplied by the rating, which is determined by a scale set in advance. In this example, the scale for the time from initial sales contact to close is 5 for a two-week cycle; 4 for a three- to six-week cycle, 3 for a seven- to nine-week cycle, 2 for a 10- to 12-week cycle, and 1 for anything beyond 12 weeks. A potential sale that would generate more than $50,000 is rated at 5, $40,000 at 4, $25,000-$40,00 at 3, $10,000-$25,000 at 2, and anything less than $10,000 would be a 1. Similarly, the likelihood of success potential criteria is 5 for greater than 90%, 4 for 80 - 90% chance, 3 for 70 - 80% chance, 2 for 60 -70% chance, and 1 for less than 60% chance.

Choosing Prospects To Pursue

	Factor (Weight)		
Prospects	**Cycle Time To Close** (1.0)	**High $$$ Potential** (1.50)	**Likelihood Of Success** (2.5)
A+ Graphics Company	3	4	4
Superior Printing	2	5	2
Fieldstone, Inc.	4	3	3

5 = ≤ 2 wks	**5 = > $50K**	**5 = 90%+**
4 = 3-6 wks	**4 = $40-$50K**	**4 = 80%-90%**
3 = 7-9 wks	**3 = $25-$45K**	**3 = 70%-80%**
2 = 10-12 wks	**2 = $10-$25K**	**2 = 60%-70%**
1 = > 12 wks	**1 = < $10K**	**1 = < 60%**

A+ Graphics: **3 (1) + 4 (1.50) + 4 (2.5) = 19**
Fieldstone, Inc.: **4 (1) + 3 (1.50) + 3 (2.5) = 16**
Superior Printing: **2 (1) + 5 (1.50) + 2 (2.5) = 14.5**

The Dogwood Company decided to pursue A+ Graphics Company first because of the apparent greater Likelihood of Success and High Dollar Volume Potential.

Questions To Ask

- Have you identified the right options to consider?
- Are the criteria right? Do you have operational definitions of them?

- Have the criteria been properly weighted? Do the weights reflect their relative value?
- Have you clearly defined the rating scale for each criterion?
- Have you arrived at a rating for each item by consensus and using data when available?
- Have you reviewed the results to determine if it makes sense? If not, it is possible that you forgot an important criterion or that the weights do not accurately reflect what is important to you.

Points To Remember

- The prioritization matrix helps you decide what to work on.
- The matrix makes the logic and decision-making visible.
- Do a "gut check" to ensure that your decision making reflects common sense.

Affinity Diagram

The affinity diagram is a tool for taking brainstormed ideas and organizing them into groups with similar themes. It is an excellent tool for creative problem-solving and for planning.

The affinity process groups similar items and then provides a theme or "header" for each group of ideas. This process helps to:

- Sort a large number of ideas quickly
- Organize random thoughts into logical groupings, while retaining the integrity of the original ideas
- Build consensus and avoid prolonged arguments
- Make logic visible
- Stimulate independent thought and a greater variety of ideas
- Make it difficult for someone to totally dominate the "discussion"

Steps In Constructing An Affinity Diagram

- Write a question about an issue you want to explore. Express it in general terms.
- Brainstorm and record the ideas about the issue on cards or Post-it notes, one idea per card.
- Place the Post-its on the wall, table, or flip chart—randomly.
- Sort the Post-its silently. Look for relationships among the ideas. Work independently, moving the cards into groupings. Sometimes, cards belong in more than one grouping. If so, make a duplicate.
- Develop header or theme cards that capture the thread of ideas for each grouping. Headers are most effective when they are written as clear and concise phrases including a noun and a verb. Place the header card at the top of the column or grouping and put a box around it to distinguish the header from other ideas.
- Review the affinity. Decide what you are going to do with this information.

Here is one team's Affinity Diagram on "Barriers to Increasing Sales Growth."

Product/Service Development Cycle Takes Too Long

- Product development delays
- Need new products and services

Sales People Need Help

- Lack of good leads
- Follow-up process not available
- Sales training too infrequent
- Salespeople don't know how to work accounts
- Need a better sales book
- Salespeople sometimes not responsive
- Need better salespeople
- Sales cycle too long
- Need more salespeople
- Need more ongoing sales from current customers

Lack Effective Customer Response

- Lack referrals from current customers
- Lack understanding of who is our customer
- We are difficult to do business with
- Customer service not good
- Delay in responding to customer complaints

Prices Not Competitive

- Price too high
- Excessive costs
- Large volume discounts not available

Product/Service Quality Needs Improvement

- Too many failures in the field
- Need better products/services to sell
- Shipping errors cause customer complaints

Senior Management Needs Focus On This Issue

- No help from senior management
- Lack of senior management interaction with customers

Questions To Ask:

- Do all of the ideas answer the question at hand?
- Do all of the ideas in each grouping share a common theme?
- Do the headers for the groupings accurately capture the theme of the ideas?
- Are there any obvious categories of ideas that have been missed in the brainstorming?
- What is the next step to take advantage of this information?
- Do the ideas and groupings point to the need for data collection?

Points To Remember

- Write a clear and concise question to be answered.
- Use formal brainstorming to generate ideas.
- Enter one idea per card to allow ideas to be easily sorted.
- All ideas should be directed at the theme or question at hand.
- Silently sort ideas into logical or creative groupings.
- Develop headers that accurately capture the theme of each group of ideas.
- Decide how this information can be best used to support your improvement efforts.

Moving Into Action

Chasing the waste out of sales is essential for success in today's competitive global economy. Only by eliminating waste—waste from lost opportunities, waste of time, waste of money, and waste of material—can you add value for customers and increase sales and gross margin dollars while reducing direct costs and expenses.

You may be uncertain about where to start but keep in mind the saying, "A long journey begins with the first step." Use this handbook as your guide, referring to the methodology as your roadmap and the concepts and tools as your vehicle. The worksheets can provide excellent starting points for moving you and your organization into action. You'll find it useful to refer back to this book throughout your continuous improvement efforts.

Remember that the keys to successfully applying the concepts of continuous improvement to sales and gross margin $ are:

- Work on the right things

- Base your decisions on facts and data
  - Data from in-depth knowledge about your work, your customers, your market, and your competitors
  - Data that tells you what the causes of problems are so you can eliminate them
  - Data that enables you to uncover opportunities for increased sales and gross margin

- Involve the people who do the work in identifying and exploiting opportunities

- Keep your focus on adding value for customers as a means to growing and improving your business

- Attack the waste with urgency!

As with any worthwhile endeavor, continuous improvement requires commitment, patience, and perseverance. But the rewards outweigh the effort. You'll find that by managing causes, you'll better manage results—results you'll be able to see on the bottom line.

Keep your eye on the prize! Once you get your first efforts underway and see how effective this approach is in increasing sales and gross margin dollars and reducing costs, you'll immediately accelerate the momentum you need to reach your goals. And once you begin chasing the waste out of sales, you'll never want to go back to business as usual.